Mid-Cretaceous Ostracoda of northeastern Algeria

STEFAN MAJORAN

Majoran, Stefan 1989 11 30: Mid-Cretaceous Ostracoda of northeastern Algeria. *Fossils and Strata*, No. 27, 1–67. Oslo. ISSN 0300-9491. ISBN 82-00-37426-2.

Algerian ostracods of the Middle Cretaceous (Albian–Cenomanian) are described and investigated with respect to their geographical and stratigraphical distributions. Access to Cenomanian assemblages of Tunisia and Israel has facilitated the investigation. It is shown that the Algerian ostracod association has its strongest affinities with other assemblages of North Africa and the Middle East, whereas there are almost no species in common with the Cenomanian assemblages of southwestern Europe. Fifty-eight species are dealt with, ten of which are new. A new genus, *Algeriana*, is erected for two of the new species, *Algeriana reymenti* sp. nov. and *Algeriana cenomanica* sp. nov.. The Cenomanian is mainly characterized by an association of *Cytherella*, *Cythereis*, *Bairdia*, *Algeriana*, 'Veeniacythereis', *Centrocythere*, *Paracypris*, *Eucythere?*, *Maghrebeis*, *Peloriops*, *Amphicytherura*, *Cytheropteron?*, *Eocytheropteron?*, *Bythocypris?*, *Pterygocythere*, *Spinoleberis* and 'Dolocytheridea'. □ *Ostracoda, Cretaceous, Albian, Cenomanian, Tethys, Maghreb, N3145 N3552 E0455 E3515.*

Stefan Majoran, Paleontologiska Institutionen, Uppsala Universitet, Box 558, S-751 22, Uppsala, Sweden; 1989 01 15 (revised 1989 05 15).

Contents

Introduction

Investigations during the last two decades have confirmed the usefulness of fossil ostracods, not only to biostratigraphers, but also to palaeoecologists. In this connection, ostracods have proved to be useful in petroleum exploration.

The present research is directed mainly to the study of Algerian ostracod associations of the Mid-Cretaceous (Albian–Cenomanian), in order to compare and chart differences with other Tethyan associations. The main part of the paper comprises taxonomic descriptions appended with stratigraphical information and details concerning palaeobiogeography.

Previously published work treating marine Cretaceous ostracods of the Maghreb (i.e. the region comprising Morocco, Algeria and Tunisia) are few, particularly if only Albian–Cenomanian forms are considered. Apart from some initial notes on Mid-Cretaceous ostracods from Algeria in Cheylan *et al.* (1953), a paper by Glintzboeckel & Magné (1959) mainly dealing with the provincial aspect of ostracods and planktonic foraminifers, serves as the basis for subsequent ostracod investigations from the southern reaches of the Tethyan sea. Some of the species illustrated by them (all given under open nomenclature) have been recorded from as far away as Kuwait (Al-Abdul-Razzaq 1977), Iran (Grosdidier 1973) and Somalia–Ethiopia (Athersuch 1988).

Using Algerian ostracods as a basis, certain large-scale stratigraphical and palaeobiogeographical properties of Cretaceous taxa were summarized by Grekoff (1969). Bassoullet & Damotte (1969) reported on Cenomanian–Turonian species from West Algeria, some of which are identical to forms illustrated by Glintzboeckel & Magné (1959).

Important contributions to our knowledge of Mid- and Upper Cretaceous ostracods of the Maghreb have been given by Bismuth *et al.* (1981a) in a monograph treating microstratigraphical and sedimentological aspects of Djebel Semmama, Tunisia. The stratigraphical sequence studied (which encompasses Upper Albian to Mid-Turonian) includes an account of the vertical distribution of about 40 species of ostracods and foraminifers respectively. Most of these ostracods had already been figured from central Tunisia by Ben Youssef (1980) and later reillustrated by Gargouri-Razgallah (1983). The latter author made use of

the ostracods, not only for stratigraphical purposes, but also in an attempt to elucidate the palaeoenvironment.

By far the most comprehensive study on marine Cretaceous ostracods (Late Albian to Campanian) of the Maghreb is that of Vivière (1985), who gave a detailed account of various aspects concerning the stratigraphy, palaeogeography and palaeoecology of species recorded in the Tebessa region, northeastern Algeria. This thesis has not been published, unfortunately, and the new taxonomic names cited therein are consequently not available under the *International Code of Zoological Nomenclature*. To enable comparisons with Vivière's work, they have nevertheless been cited in the present paper, but with clear indications that they are *nomina nuda*.

As has been pointed out earlier, for example, by Babinot (1985), Damotte (1985) and Reyment (1984), and as now supported by this investigation, there is practically no resemblance between the ostracod associations of southwestern Europe and the Maghreb at the species level, at least during the Cenomanian. There is, however, a distinct resemblance between the Algero-Tunisian associations and other associations from the southern part of the Tethys, e.g. Egypt (Colin & El Dakkak 1975), Israel (Rosenfeld & Raab 1974), Kuwait (Al-Abdul-Razzaq 1977) and Jordan (Babinot & Basha 1985). Much of this has been summarized by Athersuch (1988), Babinot (1985), Babinot & Colin (1988), Damotte (1985) and Vivière (1985).

In connection with a field-course in Israel in January 1987, I had an opportunity to collect comparative samples for this investigation through the kind help of Israeli scientists from the Geological Survey of Israel (GSI), Jerusalem; moreover, Dr. A. Rosenfeld (GSI) kindly made his personal collections available for study.

Spot samples, collected during 1983 in Djebel Semmama, Tunisia, by Dr. J. Aranki, Uppsala, have also been supplied for comparison.

Acknowledgements. – I am grateful to Dr. K.G. McKenzie, Melbourne, and Professor R.A. Reyment, Uppsala, for extensive editing and constructive criticism of drafts of the manuscript. Dr. Stefan Bengtson, Uppsala, edited and gave valuable remarks on the manuscript. Part of the SEM-work presented has been carried out by Mrs. E.R. Reyment, Uppsala. Thanks are also due to the technical staff of the Institute of Historical Geology and Palaeontology, Uppsala, for help in various ways.

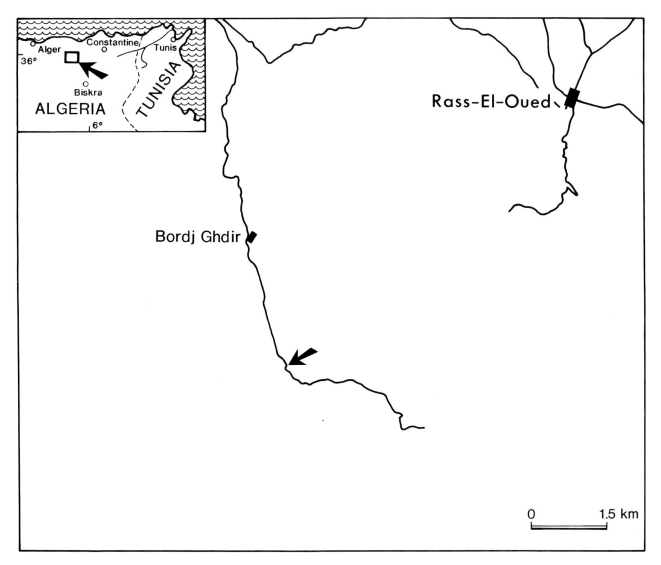

Fig. 1. Map showing the location of the Bordj Ghdir section (indicated by arrow).

Methods

The samples were broken into pieces about 2–3 cm in size, placed in a stable heat-resistant bowl, and allowed to dry in an oven (100°C) for a few hours. A saturated solution of crystallized sodium sulphate was added to the dried pieces which once again were placed in the oven (100°C) for 5–6 hours, then transferred into a refrigerator and allowed to stay there for at least 24 hours. The disintegrated samples were then covered with boiling water for a few minutes prior to washing and sieving. The sieves were passed through a solution of methylene blue in order to mark possible sources of contamination. SEM-studies of gold coated specimens were made using a JEOL T330. All figured specimens are housed in the Palaeontological Museum of Uppsala; the storage codes are PMAL for Algeria, PMTN for Tunisia, and PMIsr for Israel.

Provenance of the material

Road-section south of Bordj Ghdir

Most of the Albian–Cenomanian ostracods treated in this monograph have been obtained from a road-section some 5 km south of the village of Bordj Ghdir (lat. 35°52'N, long. 4°55'E; see Figs. 1–2), northeastern Algeria. The material was sampled in 1967 by Professor P. Marks, Utrecht, and later kindly donated to the Palaeontological Museum of Uppsala by him.

The road-section is situated on the north side of the E–W directed anticline that forms the mountain chain of Hodna. A detailed geological account of the lithology, stratigraphy and tectonics of this massif and the surrounding territory to the north, i.e. the Bordj Ghdir Basin, the Djebel M'Zaita horst and the Ras-El-Oued basin, is available in a monograph by Cruys (1955). This region is known to expose deposits from Upper Jurassic to Quaternary. Lithological and stratigraphical data for the Albian–Cenomanian in Cruys' (1955) monograph were summarized by Majoran (1987b).

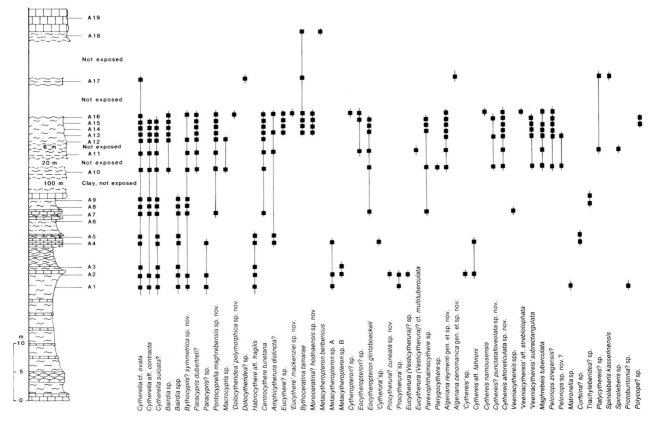

Fig. 2. Stratigraphy and ostracod ranges recorded from the Bordj Ghdir section.

The section lies practically at right angles to the strike of the Albian–Cenomanian formation (P. Marks, personal communication). Nineteen levels were sampled over a range extending from Mid-Albian to the base of the Turonian (Fig. 2).

Sample-levels A1–A9

Nine samples are from a 20 m, Middle to Upper Albian sequence consisting of dark, stiff marls alternating with layers of nodular, zoogenic limestone (Cruys 1955). On the basis of sedimentological and faunal aspects, Cruys (1955) interpreted the entire Albian sequence as originally neritic. Ostracods are not very abundant in the beds sampled here, nor are foraminifers. The ostracod species are listed below in relation to each sampling-level, given in ascending order.

A1. – Stiff, nearly black marl with the following species: *Cytherella* cf. *ovata* Roemer, *C.* aff. *contracta* Van Veen, *C. sulcata?* Van Veen, *Bairdia* spp., *Bythocypris? symmetrica* sp. nov., *Paracypris?* sp., *Habrocythere* aff. *fragilis* Triebel, *Metacytheropteron* sp. A, '*Procytherura*' sp., *Matronella* sp. and *Protobuntonia?* sp.

A2. – Same bed as A1 with the following species: *Cytherella* cf. *ovata*, *C.* aff. *contracta*, *C. sulcata?*, *Bairdia* spp., *Bythocypris? symmetrica* sp. nov., *Paracypris?* sp., *Habrocythere* aff. *fragilis*, *Metacytheropteron* sp. A, *Metacytheropteron* sp. B, *Procytherura? cuneata* sp. nov., '*Procytherura*' sp., *Eucytherura* (*Vesticytherura*)? sp., '*Cythereis*' sp. and *Cythereis* aff. *fahrioni* Bischoff.

A3. – Marl and nodular limestone with the following species: *Cytherella* cf. *ovata*, *C. sulcata?* *Bairdia* spp., *Habrocythere* aff. *fragilis* and *Metacytheropteron* sp. B.

A4. – *Terebratula* bed with the following species: *Cytherella* cf. *ovata*, *C. sulcata?*, *Bairdia* spp., *Paracypris?* sp., *Habrocythere* aff. *fragilis*, *Amphicytherura distincta?* Gerry & Rosenfeld, *Metacytheropteron* sp. A, '*Cytherura*' sp., *Cythereis* aff. *fahrioni* and *Curfsina?* sp.

A5. – Stiff, dark marl with the following species: *Cytherella* cf. *ovata*, *C. sulcata?*, *Bairdia* spp., *Habrocythere* aff. *fragilis*, *Amphicytherura distincta?* and *Curfsina?* sp.

A6. – Stiff blue marl, barren of ostracods.

A7. – Stiff, dark marl with the following species: *Cytherella* cf. *ovata*, *C.* aff. *contracta*, *C. sulcata?*, *Bairdia* spp., *Bythocypris? symmetrica*, *Pontocyprella maghrebensis* sp. nov., *Centrocythere tunetana* Bismuth & Donze, *Eocytheropteron glintzboeckeli* Donze & Lefèvre, *Parexophthalmocythere* sp. and *Veeniacythereis?* spp.

A8. – Stiff, calcareous marl, blue-gray, with the following species: *Cytherella* cf. *ovata*, *C.* aff. *contracta*, *C. sulcata?*, *Bairdia* spp., *Bythocypris? symmetrica* and *Trachyleberidea?* sp.

A9. – Same bed as A8 with the following species: *Cytherella* cf. *ovata*, *C.* aff. *contracta*, *C. sulcata?*, *Bairdia* spp., *Bythocypris? symmetrica* and *Trachyleberidea?* sp.

Sample-levels A10–A16

There is a considerable non-exposed sequence between levels A9 and A10, the latter marking the appearance of what is here considered as a typically Cenomanian ostracod association. The succession encompassed by levels A10–A16 consists of greenish-grey, marly clay. Ostracods and foraminifers are very abundant throughout, thus reflecting the conditions presented in Cruys (1955) for the Lower Cenomanian. As seen in Fig. 2, there are other non-exposed sequences between levels A10 and A11, and also between A11 and A12.

A specimen count of the 0.125–1 mm fractions was carried out at each level in order to estimate the faunal composition of the ostracods. The vast majority of these specimens were single valves. The number of individuals counted for each level varies between 200 and 1000 (see below). The species are listed in order of relative abundance with the following abundancy designations (*** indicates a relative abundance of >30%, ** 15–30%, * 5–15%; no asterisk indicates <5 %):

A10. – Number of individuals 318. *Pontocyprella maghrebensis****, *Centrocythere tunetana***, *Pterygocythere* sp.*, '*Veeniacythereis*' *subrectangulata* Majoran*, *Cythereis afroreticulata* sp. nov.*, *Cytherella* cf. *ovata*, *C.* aff. *contracta*, *C. sulcata*?, *Bairdia* sp., *Paracypris dubertreti*? Damotte & Saint-Marc, *Macrocypris* sp, *Eocytheropteron glintzboeckeli*, *Parexophthalmocythere* sp., *Bythocypris*? *symmetrica*, *Algeriana reymenti* gen. et sp. nov., *Cythereis*? *punctatafoveolata* sp. nov., *Maghrebeis tuberculata* Majoran, *Peloriops ziregensis*? (Bassoullet & Damotte), *Spinoleberis* sp. and *Peloriops* sp. nov?.

A11. – Number of individuals 223. *Cytherella* cf. *ovata****, *Pontocyprella maghrebensis***, *Paracypris dubertreti*?*, *Cytherella sulcata*?*, *Centrocythere tunetana**, *Platycythereis*? sp.*, *Cytherella* aff. *contracta*, *Bythocypris*? *symmetrica*, *Amphicytherura distincta*?, *Eocytheropteron*? sp, *Eocytheropteron glintzboeckeli*, *Eucytherura* (*Vesticytherura*)? cf. *multituberculata* Gründel, *Algeriana reymenti*, *Maghrebeis tuberculata* and *Cythereis afroreticulata*.

A12. – Number of individuals 688. *Cytherella* cf. *ovata****, *Pontocyprella maghrebensis***, *Algeriana reymenti***, *Cytherella sulcata*?*, *Peloriops ziregensis*?*, *Centrocythere tunetana*, *Cytherella* aff. *contracta*, *Bairdia* sp., *Bythocypris*? *symmetrica*, *Paracypris dubertreti*?, *Cythereis*? *punctatafoveolata*, *Cythereis afroreticulata*, '*Veeniacythereis*' *subrectangulata*, *Maghrebeis tuberculata*, *Peloriops* sp. nov?, '*Veeniacytheris*' sp. nov?, and *Macrocypris* sp..

A13. – Number of individuals 582. *Cytherella* cf. *ovata***, *Pontocyprella maghrebensis***, *Cythereis afroreticulata***, *Bairdia* sp. , *Algeriana reymenti**, *Cytherella* aff. *contracta*, *C. sulcata*?, *Paracypris dubertreti*?, *Centrocythere tunetana*, *Amphicytherura distincta*?, *Bythoceratina tamarae*, *Monoceratina*? *hodnaensis* sp. nov., *Eocytheropteron glintzboeckeli*, *Parexophthalmocythere* sp., '*Veeniacythereis*' *subrectangulata*, *Maghrebeis tuberculata*, *Peloriops ziregensis*? and *Eucythere*? sp.

A14. – Number of individuals 978. *Bairdia* sp.***, *Pontocyprella maghrebensis***, *Algeriana reymenti***, *Cytherella* cf.

ovata**, *Cythereis afroreticulata**, *Cytherella* aff. *contracta*, *C. sulcata*?, *Paracypris dubertreti*?, *Centrocythere tunetana*, *Eucythere*? sp., *Bythoceratina tamarae*, *Monoceratina*? *hodnaensis*, *Eocytheropteron glintzboeckeli*, *Parexophthalmocythere* sp., '*Veeniacythe reis*' *subrectangulata*, *Maghrebeis tuberculata*, *Peloriops ziregensis*? and *Polycope*? sp.

A15. – Number of individuals 526. *Cythereis afroreticulata****, *Pontocyprella maghrebensis***, *Cytherella sulcata*?*, '*Veeniacythereis*' *subrectangulata**, *Centrocythere tunetana**, *Cytherella* cf. *ovata*, *C.* aff. *contracta*, *Bairdia* sp., *Paracypris dubertreti*?, *Eucythere*? sp., *Bythoceratina tamarae*, *Monoceratina*? *hodnaensis*, *Eocytheropteron glintzboeckeli*, *Eocytheropteron*? sp., *Parexophthalmocythere* sp., *Algeriana reymenti*, *Cythereis fahrioni*, *Cythereis*? *punctatafoveolata*, *Peloriops ziregensis*? and *Polycope*? sp.

A16. – Number of individuals 753. *Cytherella* cf. *ovata***, *Cythereis afroreticulata**, *Maghrebeis tuberculata**, *Pontocyprella maghrebensis**, '*Dolocytheridea*' *polymorphica* sp. nov.*, *Algeriana reymenti**, *Cytherella* aff. *contracta*, *Bythocypris*? *symmetrica*, *Bairdia* sp., *Paracypris dubertreti*?, *Centrocythere tunetana*, *Amphicytherura distincta*?, *Eucythere*? sp., *Bythoceratina tamarae*, *Monoceratina*? *hodnaensis*, *Cytheropteron*? sp., *Eocytheropteron*? sp., *Cythereis namousensis*, *Peloriops ziregensis*?, '*Eucythere*' *mackenziei* sp. nov. and '*Veeniacythereis*' aff. *streblolophata*.

Environmental implications of the A10–A16 associations. – The large and more abundant ostracods of this section (e.g., *Cytherella* cf. *ovata*, *Pontocyprella maghrebensis*, *Algeriana reymenti*, *Cythereis afroretiulata* and *Bairdia* sp.), display distributional patterns which accord quite well with the criteria established by Whatley (1983) for an autochthonous ostracod association in a rather low energy biocoenosis. Instars of the A-4 stage are common among these forms. The only larger, abundant species in this sequence that entirely lacks juveniles is '*Veeniacythereis*' *subrectangulata*.

Each level in this part of the section, except for level A16, is characterized by a dominating fraction of two to three species that constitute as much as 50–80% of all individuals. In relation to *Pontocyprella maghrebensis*, always one of these dominating species, this may point towards an open marine environment beyond the edge of the carbonate platform (cf. Babinot & Colin 1983), or possibly an association to an outer sea-shelf. Similar indications are manifested particularly at levels A11–A14 by the great abundance of *Cytherella* cf. *ovata*, especially when it is associated with abundant *Bairdia* sp. (levels A13–A14). These two species exhibit the characteristic open marine properties displayed by representatives from southwestern Europe (cf. Babinot & Colin 1983). In this connection, one has to note that strongly sulcate representatives of *Cytherella*, e.g. *Cytherella gigantosulcata* Rosenfeld, are generally associated with inner platform environments on both sides of the Tethyan sea (cf. Babinot & Colin 1983; Vivière 1985). In order to check these environmental implications, a P/B-ratio estimation of the foraminifers was carried out in two of the samples (A14 and A15) (P/B-ratio = the quotient between the number of planktonic individuals and the total number of individuals, benthic as well as planktonic). A subsample of the 125–500 µm fraction of each sample was taken using a modified Otto microsplitter, and approximately 300 individuals were counted in each sample. Planktonic forms make up 74% and 79%, respectively, of the total foraminiferal content. This independent indication of offshore conditions (cf. Hart 1980; Vivière 1985) accords well with the conclusions based on ostracods.

It is interesting to note that two of the more abundant ostracod species, *Algeriana reymenti* and *Cythereis afroreticulata*, are strongly ornate and thick-shelled (i.e. they display characters assumed to have evolved in a shallow-water, high-energy environment according to Whatley 1983). In the present case, however, they are associated with *Pontocyprella*, and hence a fairly low energy environment is indicated. This is perhaps not an unusual coincidence, and one can at least infer that the depth is less than 800 m since these trachyleberidids exhibit well-developed eye-tubercles (cf. Benson 1975). Anoxia may have prevailed, at least occasionally, during parts of the sequence, as intimated by a significant number of pyritized *Cytherella* cf. *ovata* valves in levels A12 and A13.

Finally one needs to question whether the implications based on data from southwestern Europe (Babinot & Colin 1983) are entirely apposite to explain contemporary conditions in North Africa. Little is yet known about the specific environmental conditions that prevailed along the south reaches of the Tethys, and more aspects ought to be explored before definite conclusions may be reached.

Sample-levels A17–A18

There is a non-exposed sequence between levels A16 and A17, also between levels A17 and A18. Ostracods and foraminifers are not abundant in the two samples, nor are they well preserved. Apart from a dozen carapaces of *Dolocytheridea* sp., the species listed below are represented by only a few individuals. Cruys (1955, p. 234) considered that the environment in which the uppermost Cenomanian marls (level A18) were deposited was neritic.

A17. – Marly clay (yellow) with the following species: *Cytherella* cf. *ovata*, *Dolocytheridea?* sp., *Bythoceratina tamarae*, *Algeriana cenomanica*, *Platycythereis?* sp. and *Spinoleberis kasserinensis* Bismuth & Saint-Marc.

A18. – Yellow marl with the following species: *Bythoceratina tamarae* and *Metacytheropteron berbericus* (Bassoullet & Damotte).

Sample-level A19

Bed of Turonian rudist and algal limestone. Barren of ostracods.

Djebel Semmama (Tunisia)

This composite section ('DZ') is well documented in a previous detailed sedimentological and biostratigraphical account of Djebel Semmama (Bismuth *et al.* 1981a, 1982).

Comparative samples for this particular investigation were obtained from the lower (predominantly marly) part of the 'Ben Younes' sequence, i.e. from the Lower Cenomanian (Bismuth *et al.* 1981a, p. 200). On the basis of the

microfaunal assemblages in association with coquina and oysters, Bismuth *et al.* (1982, p. 176) interpreted this unit as originally having been deposited in a relatively shallow marine environment.

The following species were obtained for comparison here: *Centrocythere tunetana*, *Bythoceratina tamarae*, *Eocytheropteron glintzboeckeli*, *Algeriana cenomanica*, *Cythereis fahrioni bigrandis* subsp. nov., *Peloriops ziregensis?* and *Spinoleberis kasserinensis*.

Qastel, Jerusalem (Israel)

This section, which is situated outside Jerusalem along the road to Tel-Aviv (coord. 16377/13391), is part of the Moza Formation (Arkin *et al.* 1965). A detailed chart of its lithology and ostracod content is given in Rosenfeld & Raab (1974, p. 52). The section comprises mainly the Upper Cenomanian UC-3 assemblage zone of Rosenfeld & Raab (1974) which according to them was deposited in a shallow marine environment.

The following species, sampled from the UC-3 assemblage zone, was obtained for comparison in this paper: *Bairdia* sp., *Amphicytherura distincta?* *Metacytheropteron berbericus*, *Cythereis namousensis* and *Peloriops* sp.

Entrance to Hamakhtesh Hagadol (Israel)

The Cenomanian part of this road-section (coord. 15270/04189), has been charted in detail with respect to its lithology and ostracod content by Rosenfeld & Raab (1974, pp. 38–39). Six samples were obtained from the Lower Cenomanian Hevyon member, which belongs to the lower part of the Hazera formation (Arkin & Braun 1965). Only one of the samples was reasonably rich in ostracods, yielding mainly *Cytherella* spp., also a few *Dolocytheridea?* sp., '*Veeniacythereis*' sp. and *Cythereis namousensis*.

Systematics

The systematic framework is a simplified version of Hartmann & Puri (1974) in that many subfamilies and all tribes are omitted. Many species remain in open nomenclature; the meaning of quotation marks, 'aff.', 'cf.' and '?' follows the recommendations of Bengtson (1988). Terminology concerning ornamental details, particularly as applied to the trachyleberidids, follows Sylvester-Bradley & Benson (1971). In the systematics below, the information following 'occurrence' concerns vertical distribution in relation to the investigated samples only, whereas the information following 'distribution' summarizes what is presently known about the geographical and stratigraphical distribution of the species. The road-section south of Bordj Ghdir will henceforward be referred to as the 'Bordj Ghdir section'.

Order Podocopida Müller, 1894

Suborder Platycopa Sars, 1866

Family Cytherellidae Sars, 1866

Genus *Cytherella* Jones, 1849

Remarks. – One of the first problems one faces when dealing with an abundance of cytherellids, is to try to separate the different species. This is not always an easy task, since shape, which often serves as the only available basis for separation, may turn out to be polymorphic (cf. Ducasse 1981). However, once a separation of the different species has been accomplished, next comes their identification. In this connection, one cannot possibly avoid nor easily solve the problems summarized by Reyment (1984, p. 74).

The few cytherellid species of the present collection, however, seem reasonably distinct and display also a strong affinity to already named taxa.

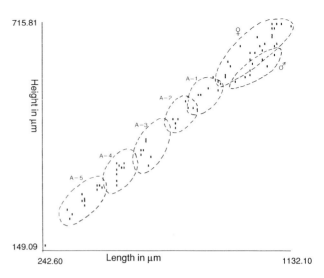

Fig. 3. Length–height plot for *Cytherella* cf. *ovata* Roemer. (*N*=77 right valves.)

Cytherella cf. *ovata* Roemer
Pl. 1:1–8

Synonymy. – See Babinot 1980.

Material. – Hundreds of valves and carapaces.

Occurrence. – Bordj Ghdir section, levels A1–A17.

Measurements. – See the ontogenetic distribution in Fig. 3.

Description. – The carapace is smooth. The individuals regarded as females are ovoid with subconvex ventral and convex dorsal margins. The males are more elongate subrectangular with nearly straight, subparallel anterodorsal and ventral margins. A significant feature is the sharply angulated, posterodorsal periphery which, however, tends to lose some of its distinctiveness among the females. Greatest height is at mid-length, at the anterior end of this angulated section. The anterior margin is broad and regularly rounded, the posterior somewhat more narrowly so. Greatest width is slightly posterior of the mid-length. In dorsal view, males taper almost equally towards the anterior and posterior, whereas females are more inflated posteriorly.

Remarks. – The larger individuals (adults) of what is believed to be a single species are figured. Most of the size and shape variation can be explained as being due to sexual dimorphism; however, certain variation with respect to the outline can also be followed back through the last 3–4 ontogenetic stages. There seems also to be a significant increase in average size from the Albian to the Cenomanian, albeit this is partly blurred by difficulties in distinguishing between the adult and the A-1 stage. If we are really dealing with a single species, the lack of distinct ontogenetic clusters (Fig. 3) suggests that the species has a good tolerance to environmental fluctuations (cf. Szczechura 1971). Another possibility is that there were several generations per season (cf. Abe 1983).

Apart from some deviation with respect to the anterior outline, there is a strong resemblance to the 'locotypes' of *Cytherella ovata* as figured by Damotte (1971).

Cytherella aff. *contracta* Van Veen
Pl. 1:9–10

Synonymy. – □aff. 1932 *Cytherella contracta* nov. spec. – Van Veen, pp. 343–344, Pls. 8, 9:1–4.

Material. – About 30 carapaces and valves.

Occurrence. – Bordj Ghdir section, levels A1–A2, A7–A15.

Measurements. – Length 0.59–0.64 mm, height 0.31–0.33 mm.

Remarks. – The carapace is smooth and elongate subrectangular, with weakly sinuous, subparallel dorsal and ventral margins. The anterior margin is semicircular, the posterior somewhat more oblique. With respect to the lateral outline, there is a strong resemblance to *Cytherella contracta* Van Veen, although in dorsal view the carapace seems to terminate more sharply. The greatest width is slightly posterior to the mid-length.

The Bordj Ghdir section has also yielded a few similar individuals of larger size (length 0.77 mm, height 0.41 mm).

Cytherella sulcata? Van Veen
Pls. 1:11–15; 17:15–16

Synonymy. – □?1932 *Cytherella sulcata* nov. spec. – Van Veen, pp. 336–337, Pl. 4.

Material. – About 200 carapaces and valves.

Occurrence. – Bordj Ghdir section, levels A1–A15.

Measurements. – Length 0.62–0.72 mm, height 0.38–0.44 mm.

Description. – The carapace is smooth, elongate subrectangular, and displays a convex dorsal margin which is subparallel to the nearly straight to weakly convex ventral margin. There is, however, a slight indentation at the midlength of the left valve dorsal margin just above the rather significant but weak dorsomedian sulcus. The anterior and posterior margins are semi-circular. In dorsal view, the outline is wedge-shaped, widest near the rear and tapering towards the anterior; a few individuals, presumably males, are more compressed posteriorly than those figured here.

Remarks. – The morphology is surprisingly similar to the types of the Maastrichtian *Cytherella sulcata* (cf. Van Veen 1932) from which they cannot readily be differentiated, despite the considerable difference with respect to stratigraphical occurrence. Cytherellids are, however, in some cases associated with an immense stratigraphical range, e.g. *Cytherella ovata*.

The individuals figured on Pl. 1:14–15 are rather exceptional in that the posterior inflation and the dorsal convexity are exaggerated.

Superfamily Bairdiacea Sars, 1866

Family Bairdiidae Sars, 1888

Genus *Bairdia* McCoy, 1844

Bairdia sp.
Pl. 2:5–7

Material. – Hundreds of valves and carapaces, including juveniles.

Occurrence. – Bordj Ghdir section, levels A10, A12–A16.

Measurements. – Length 1.03–1.08 mm, height 0.68–0.70 mm.

Description. – The valves of this comparatively large representative of the genus are strongly calcified, smooth and unequal. The left valve overreaches the entire circumference of the right one, especially mid-dorsally and mid-ventrally. The left valve displays a highly arched mid-dorsal margin (greatest height at mid-length) and straightened antero- and posterodorsal margins; the latter terminates in a salient cauda. The ventral margin of this valve is straight at mid-length but curves upward posteriorly and more strongly so anteriorly to form the broadly rounded anterior.

The dorsal margin of the right valve is tripartite; all segments are straight, although this is less clear on the individual figured here. The posterior is distinctly caudate. The ventral margin is slightly inflexed at mid-length, but agrees otherwise in outline with the left valve. Some individuals of both valves show small denticles along the postero-ventral periphery. The anterior is somewhat narrower and more oblique than in the left valve.

In dorsal view, the carapace is fusiform and moderately expanded with sharp ends. The valves are recrystallized. 'Bairdoppilatid' teeth or sockets are not observable, neither are muscle-scars nor other finer details.

Remarks. – The gross morphology coincides quite well with that summarized in Babinot & Colin (1983, p. 186) as representative of the open marine assemblages of southwestern Europe (Upper Cretaceous). A few Israeli specimens in the comparative collection (Qastel) seem, however, even more closely comparable to the forms described (cf. Rosenfeld & Raab 1974).

Bairdia spp.
Pl. 2:1–4

Material. – About thirty valves and carapaces.

Occurrence. – Bordj Ghdir section, levels A1–A9.

Measurements. – Length 1.10–1.20 mm, height 0.62–0.67 mm.

Remarks. – The Albian part of the section has yielded mostly poorly preserved specimens briefly considered here. It is possible that more than one species is included because of some variation with respect to outline and length. Some of the best preserved individuals are figured.

Compared with the Cenomanian species discussed above, the dorsal margin of the right valve in this taxon is not as distinctly tripartite, nor is the left valve as caudate. The figured left valve lacks a cauda entirely and has a triangular outline in lateral aspect. This valve is almost symmetrical with respect to a vertical plane at mid-length, the site of greatest valve-height.

Family Bythocyprididae Maddocks, 1969

Genus *Bythocypris* Brady, 1880

Bythocypris? symmetrica sp. nov.
Pl. 2:8–9

Synonymy. – □1985 *Bythocypris* sp.1 – Vivière, p. 145, Pl. 2:12.

Holotype. – PMAL54 (Pl. 2:9). Bordj Ghdir section, level A10.

Derivation of the name. – After the symmetry with respect to a sagittal plane at mid-length.

Material. – About 30 carapaces.

Occurrence. – Bordj Ghdir section, levels A1–A2, A7–A12, A16.

Measurements. – Length 0.67–0.69 mm, height 0.31–0.32 mm.

Diagnosis. – A species of *Bythocypris?* which is almost symmetrical with respect to a sagittal plane at mid-length.

Description. – The valves are smooth and subequal, the carapace being almost symmetrical with respect to a sagittal plane at mid-length (i.e. at the site of greatest height and length). The posterior is only slightly narrower than the anterior. The dorsal margin is regularly arched, the ventral almost straight, although slightly inflexed at mid-length in the right valve. Viewed dorsally, the carapace is fusiform. The internal features were not available for study.

The Danian *'Bythocypris'* sp. (Esker 1968, p. 321) from Tunisia, coincides in gross morphology with this form.

Superfamily Cypridacea Baird, 1845

Family Candonidae Kaufmann, 1900

Genus *Paracypris* Sars, 1866

Paracypris dubertreti? Damotte & Saint-Marc
Pl. 2:10–12

Synonymy. – □?1972 *Paracypris dubertreti* n. sp. – Damotte & Saint-Marc, p. 276, Pl. 1:1. □?1974 *Paracypris acutocaudata* n. sp. – Rosenfeld; Rosenfeld & Raab, p. 8, Pl. 1:22–24. □1985 *Paracypris dubertreti* Damotte & Saint-Marc – Vivière, p. 149, Pl. 3:6–7.

Material. – About 100 carapaces and valves.

Occurrence. – Bordj Ghdir section, levels A10–A16.

Distribution. – Cenomanian (and Lower Turonian) of Algeria; Cenomanian of Lebanon. For references see synonymy above.

Measurements. – Length 0.74–0.84 mm, height 0.31–0.33 mm.

Description. – The carapace is smooth and elongate wedge- or dagger-shaped in lateral aspect. In dorsal view, it is fusiform with greatest width anterior to mid-length. The left valve is larger than the right, overlapping it most along the concave ventral margin. The anterior margin is almost evenly rounded but the posterior prolongs at an angle, approximately ⅓ of the length from the anterior end. The dorsal margin appears bipartitioned, as it is characterized by a marked posterodorsal angle between two almost straight sections. The posterior is tapered and its extremity sharply pointed. Internal features are typical for the genus.

Remarks. – The fact that the present forms are significantly smaller than the types (cf. Damotte & Saint-Marc 1972) precludes a definite assignation to *Paracypris dubertreti*.

Paracypris? sp.
Pl. 2:13–15

Material. – About 30 carapaces.

Occurrence. – Bordj Ghdir section, levels A1–A2, A4.

Measurements. – Length 0.84–0.86 mm, height 0.32–0.34 mm.

Description. – The valves are smooth and subreniform–lanceolate in lateral aspect. Greatest height occurs at the weakly expressed dorsal angle, approximately ⅓ of the length from the anterior end. There is an almost straight section anterior to that angle, whereas posteriorly the dorsal margin is regularly arched. The anterior margin is obtusely rounded, the posterior more narrowly so, or subacute, although always with its extremity level with the straight ventral margin. The left valve overreaches the right along the anterodorsal and ventral margins. Greatest width lies at mid-length as seen from its elliptical outline in dorsal view. Internal features were not observable.

Remarks. – The generic assignation is somewhat unertain, although *Paracypris mdaourensis* Bassoullet & Damotte, closely resembles the present forms. *Paracypris mdaourensis* differs, however, by its more inflated anterior region and also by the anteroventral outline (cf. Bassoullet & Damotte 1969).

Family Pontocyprididae Müller, 1894

Genus *Pontocyprella* Mandelstam *in* Ljubimova, 1955

Pontocyprella maghrebensis sp. nov.
Pl. 3:1–3

Synonymy. – □1985 *Pontocyprella* sp.1 – Vivière, pp. 146–147, Pl. 3:1.

Holotype. – PMAL63 (Pl. 3:3). Bordj Ghdir section, level A14.

Derivation of the name. – After its geographical provenance.

Material. – Hundreds of valves and carapaces.

Occurrence. – Bordj Ghdir section, levels A7?, A10–A16.

Measurements. – Length 0.77–0.87 mm, height 0.41–0.43 mm.

Diagnosis. – A species of *Pontocyprella* with left valve displaying a convex venter, high dorsum and rounded posterior.

Description. – The smooth carapace is inequivalved and bean-shaped. In dorsal view, it is subelliptical with its greatest width slightly posterior to mid-length. The left valve overlaps the entire circumference of the right one, especially mid-ventrally and anterodorsally. The dorsal margin reaches its greatest height at the junction between its substraight anterior third and more regularly arched posterior two thirds. The posterior is narrowly rounded, its extremity slightly below mid-height. The anterior margin is oblique, somewhat upturned dorsally where it forms an angle at the junction with the dorsal margin. The hinge accords with the generic assignment but other internal details are too recrystallized to yield information of significance.

Remarks. – This species accords well with *Pontocyprella* in the shape distinction made between the genera *Bythocypris* and *Pontocyprella* by Rodriguez Lazaro (1985, p. 147). There is a strong resemblance to the Danian *Pontocyprella recurva* Esker, as pointed out by Vivière (1985, p. 147), who also noted some disagreements with respect to the posterodorsal outline between the latter and the present forms.

Family Macrocyprididae Müller, 1912

Genus *Macrocypris* Brady, 1868

Macrocypris sp.
Pl. 3:4

Material. – A few poorly preserved carapaces.

Occurrence. – Bordj Ghdir section, levels A10, A12.

Measurements. – Length 0.97 mm, height 0.36 mm. (Specimen No. PMAL64).

Remarks. – The individual figured is one of a few deformed and very poorly preserved specimens with the external properties of *Macrocypris*.

Superfamily Cytheracea Baird, 1850

Family Cytherideidae Sars, 1925

Genus *Dolocytheridea* Triebel, 1938

'Dolocytheridea' polymorphica sp. nov.
Pl. 3:10–13; Pl. 17:6–14

Holotype. – PMAL71 (Pl. 3:11). Bordj Ghdir section, level A16.

Derivation of the name. – Named in reference to its polymorphism in shape.

Material. – About 40 valves.

Occurrence. – Bordj Ghdir section, level A16.

Measurements. – Length (males and females) 0.52–0.58 mm; height males 0.27–0.29 mm, females 0.25–0.27 mm.

Diagnosis. – A small species of *'Dolocytheridea'* with unequal valves strongly affected by polymorphism in shape, and with the posterior extremity level with the ventral margin.

Description. – The valves are smooth, unequal and rather inflated. There is significant dimorphic variation in size which appears sexually related; this is also clearly affected by shape polymorphism, particularly manifested in the triangular, highly arched dorsal margin of the more elongate individuals (Pl. 17:10–14), i.e. those interpreted as males. In the latter category, there is also some size variation which is wide enough to encompass the generally smaller females. The greatest width in both categories is slightly posterior to the mid-length, greatest height approximately at mid-length. The ventral margin is straight in the left valve but slightly concave in the right valve. The posterior extremity level with the ventral margin. The anterior is broadly rounded.

Remarks. – Internal features are not readily observable owing to recrystallization; this may possibly cast doubt on the erection of a new species. The specimens seem nevertheless like a well defined subset at the species level. The polymorphism is the continuous polymorphism or polyphenism of Mayr (1963), that is ecophenotypic in nature (cf. Reyment, in press).

The larger individuals appar very close to *Dolocytheridea atlasica* Bassoullet & Damotte, although that species shows no trace of di- or polymorphic patterns, according to Bassoullet & Damotte (1969).

Dolocytheridea? sp.
Pl. 3:5–9

Material. – About twenty carapaces.

Occurrence. – Bordj Ghdir section, level A17.

Measurements. – Length males 0.74–0.77 mm, females 0.67–0.69 mm. Height males 0.38 mm, females 0.36–0.38 mm.

Remarks. – Some smooth and rather inflated carapaces, exhibiting what is interpreted as pronounced sexual dimorphism, are considered here. The left valve overlaps almost the entire margin of the right valve. The shape in lateral aspect is like *Paracyprideis* or *Dolocytheridea*. The latter generic name, however, seems more appropriate as far as can be judged from the shape in dorsal view in relation to sexual dimorphism; i.e. the greatest width is posterior to mid-length in males but at mid-length in females (cf. Van Morkhoven 1963).

Males are elongate and bean-shaped with inflexed ventral and weakly sinuous anterodorsal margins. The females are shorter with an arched dorsal margin and a substraight venter. Common features are the broad anterior, the posterodorsal curvature and the shape of the posterior extremity. Internal features were not available for study.

Genus *Habrocythere* Triebel, 1940

Habrocythere aff. *fragilis* Triebel
Pl. 3:14–15

Synonymy. – □aff. 1941 *Habrocythere fragilis* n. sp. – Triebel, pp. 166–169, Pls. 1:10–13; 9:101.

Material. – A dozen carapaces.

Occurrence. – Bordj Ghdir section, levels A1–A5.

Measurements. – Length 0.45–0.47 mm, height 0.26–0.28 mm.

Remarks. – An assignment to *Habrocythere* seems to be the best choice for the specimens figured. Neale (1982) discussed the classification of *Habrocythere* and related genera in the light of internal features, but since they cannot be examined here, this determination is based on external characters only, *viz.* the size, shape and, especially, the anteromarginal compression. The latter is better displayed in the right valve, where it forms a shallow moat.

The left valve overreaches nearly the entire margin of the right one, most conspicuously at the site of greatest height and along the anteroventral margin. The outline resembles that of *Habrocythere fragilis* Triebel, except that it lacks the angulated posterodorsal margin of that species (cf. Triebel 1940). Moreover, no antero- and posteromarginal denticles are preserved, nor any surface ornament such as that which characterizes *H. fragilis*.

Family Cytheridae Baird, 1850

Genus *Centrocythere* Mertens, 1956

Centrocythere tunetana Bismuth & Donze
Pl. 4:1–5

Synonymy. – □1959 Ostracode B8 – Glintzboeckel & Magné, p. 62, Pl. 2:19. □?1974 *Centrocythere denticulata* Mertens – Rosenfeld & Raab, p. 13, Pl. 2:29–31. □1980 *Centrocythere tunetana* Bismuth & Donze – Ben Youssef, p. 87, Pl. 4:11–12. □1981b *Centrocythere tunetana* n sp. Bismuth & Donze – Bismuth *et al.*, p. 57, Pl. 2:10–11. □1981a *Centrocythere tunetana* Bismuth & Donze – Bismuth *et al.*, p. 227, Pl. 8:13–14. □1983 *Centrocythere tunetana* Bismuth & Donze – Gargouri-Razgallah, Pl. 28:1–2. □1985 *Centrocythere tunetana* Bismuth & Donze – Vivière, pp. 162–162, Pl. 5:14–15.

Material. – About hundred carapaces and valves.

Occurrence. – Bordj Ghdir section, levels A7, A10–A16. Also from Djebel Semmama (Lower Cenomanian).

Distribution. – Upper Albian – Cenomanian of Algeria and Tunisia; also the Cenomanian of Israel. For references see synonymy above.

Measurements. – Length males 0.56–0.62 mm, females 0.51–0.54 mm. Height males and females 0.27–0.29 mm.

Remarks. – This species is rather abundant in the present collection and has frequently been reported from Algeria and Tunisia (see synonymy). The carapace is very inflated and stocky, so much so that ventrally it obscures the ventral valve border. The overall punctate surface is associated with a lateroperipheral arrangement of concentric costae. There is a marked eye-tubercle. The hinge accords well with that of the genus.

To the original description (Bismuth *et al.* 1981b) can be added the observation that individuals in the Algerian collection appear distinctly dimorphic with respect to length. Another remarkable feature, shown by at least the Tunisian specimens, is the presence of strutted outgrowths on the reticulum muri.

Genus *Amphicytherura* Butler & Jones, 1957

Amphicytherura distincta? Gerry & Rosenfeld
Pl 4:6–9

Synonymy. – □?1973 *Amphicytherura distincta* n. sp. – Gerry & Rosenfeld, p. 99, Pl. 2:7–10. □?1974 *Amphicytherura distincta* Gerry & Rosenfeld – Rosenfeld & Raab, p. 16, Pl. 2:41–42. □?1980 *Amphicytherura distincta* Gerry & Rosenfeld – Ben Youssef, p. 91, Pl. 4:15–16. □1981a *Amphicytherura distincta* Gerry & Rosenfeld – Bismuth *et al.* p. 228, Pl. 7:16–18. □1983 *Amphicytherura distincta* Gerry & Rosenfeld – Gargouri-Razgallah, p. 152, Pl. 28:3–4. □1985 *Amphicytherura distincta* Gerry & Rosenfeld – Vivière, p. 255, Pl. 27:7.

Material. – About 30 carapaces and valves.

Occurrence. – Bordj Ghdir section, levels A4–A5, A11, A13, A16.

Distribution. – Cenomanian of Algeria, Tunisia and Israel. For references see synonymy above.

Measurements. – Length 0.41–0.42 mm or 0.33–0.34 mm. Height 0.25–0.26 mm or 0.21–0.22 mm.

Remarks. – Against a definite assignation to *Amphicytherura distincta*, speaks the fact that the Algerian forms, apart from displaying better expressed secondary reticula than the specimens of Gerry & Rosenfeld (1973), encompass two

size-classes, a larger for the *Albian* individuals (which roughly accords with the Upper Cenomanian forms of Gerry & Rosenfeld 1973) and a smaller category of Cenomanian forms. Moreover, the caudal process is well expressed on their forms but missing from those figured here, a consequence possibly of abrasion caused by the ultrasonic treatment.

The carapace is subrhomboidal in lateral aspect and compressed in dorsal view. There is a median ridge, which rises from front to rear and from which half a dozen minor ridges lead off perpendicularly on either side to join the dorso- and ventromarginal ridges respectively. Between these vertical ridges there are well expressed secondary reticula. The anterior ventromarginal ridge is bifurcated; the eye-tubercle is prominent.

Family Eucytheridae Puri, 1954

Genus *Eucythere* Brady, 1868

Eucythere? sp.
Pl. 4:10–15

Synonymy. – □?1985 *Eucythere ramentosa* n. sp. – Vivière, pp. 252–253, Pl. 26:11–14 [unpublished name; *nomen nudum* herein (cf. p. 3)].

Material. – About 30 carapaces and valves.

Occurrence. – Bordj Ghdir section, levels A13–A16.

Distribution. – Cenomanian of Algeria (see Vivière 1985).

Measurements. – Length 0.67–0.72 mm or 0.46–0.56 mm. Height 0.38–0.44 mm or 0.31 mm.

Remarks. – The present forms can be divided into two size-classes. The smaller forms are identical to *Eucythere ramentosa* [*nomen nudum*] (cf. Vivière 1985), whereas the larger transcend the size limits of that species significantly. The mutual relationships between the two size categories may possibly be explained in terms of size polymorphism, because the shape seems identical in the two categories. Internal features are too poorly preserved to yield any information of significance.

Description. – The carapace is triangular in lateral aspect and elliptical in dorsal and ventral views. There is a slight inflexure where the broadly rounded anterior margin joins the venter. The dorsal margin is straight and converges strongly towards the narrower posterior. The ventral margin is relatively straight. There is a broad, crescent-shaped, anteromarginal compression defined by the evenly rounded anterior margin and the marked border of the moderately inflated main region of the valve. An elongate eye-spot is vaguely visible. The valve surface is either entirely smooth (as on most larger forms) or it may display a few striate ridgelets in the ventromedian and mid-ventral region (as on most smaller forms). Small pits are visible in the mid-ventral region of the specimen figured in ventral view.

'Eucythere' mackenziei sp. nov.
Pl. 5:1–5

Holotype. – PMAL89 (Pl. 5:1). Bordj Ghdir section, level A16.

Derivation of the name. – After Dr. K.G. McKenzie, Melbourne, Australia

Material. – A dozen valves and a single juvenile? carapace.

Occurrence. – Bordj Ghdir section, level A16.

Measurements. – Length 0.46–0.49 mm, height 0.26–0.28 mm.

Diagnosis. – A species of 'Eucythere' provided with eye-spots, and with the main part of the valve inflated and reticulate; anterior and posterior regions compressed.

Description. – The triangular valves are unequal, although details concerning overlap are presently unknown. The major part of the lateral surface is inflated, ornamented with irregular reticula which are partly disrupted centrally. There is a weak eye-spot from which two tiny ridges run off downwards to the smooth and compressed anterolateral region. The posterolateral region is also smooth and equally compressed. The dorsal margin is straight and obscured by the convex dorsal portion of the valve, as the weakly sinuous ventral margin is by the convex ventral portion. The anterior margin is evenly rounded, though considerably broader in the right valve. The posterior is narrow. There are a few longitudinal ridgelets in the inflated mid-region of the ventral surface. Of the internal features, only a poorly preserved hinge could be observed; it appears to be merodont/entomodont *sensu* Van Morkhoven (1962).

Remarks. – The Bordj Ghdir section has yielded rare specimens of what is possibly a new genus. There are certain characteristics in common with *Eucythere* (*sensu* Van Morkhoven 1962, p. 337–340), particularly with respect to size and shape; however, the new forms can be readily differentiated because they exhibit eye-spots and rather well expressed lateral reticula. *Neocythere* is also comparable, but larger and more regularly ornamented. Juveniles of *Neocythere* may be indistinguishable from the new species.

Family Bythoceratiidae Sars, 1866

Genus *Bythoceratina* Hornibrook, 1952

Bythoceratina tamarae Rosenfeld
Pl. 5:6–9

Synonymy. – □1959 Ostracode D9 – Glintzboeckel & Magné, p. 64, Pl. 3:29. □1974 *Bythoceratina tamarae* n. sp. – Rosenfeld; Rosenfeld & Raab, p. 10, Pl. 2:3–5; Pl. 4:7–8. □1980 *Bythoceratina tamarae* Rosenfeld – Ben Youssef, p. 93, Pl. 5:5. □1981a *Bythoceratina tamarae* Rosenfeld – Bismuth *et al.* p. 224, Pl. 8:4–6. □1983 *Bythoceratina tamarae* Rosenfeld – Gargouri-Razgallah, p. 149, Pl. 26:4–6. □1985 *Bythoceratina tamarae* Rosenfeld – Vivière, pp. 152–153, Pl. 3:13–14.

Material. – About a dozen carapaces, no single valves.

Occurrence. – Bordj Ghdir section, levels A13–A17. Also Djebel Semmama (Lower Cenomanian).

Distribution. – Cenomanian of Algeria, Tunisia and Israel. For references see the synonymy above.

Measurements. – Length 0.64–0.65 mm, height 0.35–0.36 mm.

Description. – The dorsal and ventral margins are parallel and nearly straight, the latter being shorter than the former which terminates in a posterodorsal cauda. The valves are subequal. The lateral surface, which is almost entirely covered with a network of quadrangular reticula, displays a lobe-like swelling on each side of a distinct dorsomedian sulcus. Also prominent is the powerful, ventromedian, mamillate spine which is adorned with concentric striae. The depression of the dorsal valve border in a shallow furrow is visible in dorsal view. Internal features could not be studied.

Remarks. – A comparison between different reports on this species (e.g. Rosenfeld & Raab (1974) and Vivière (1985)), reveals certain distinct differences with respect to average size. In that respect, the present Algerian forms are very close to those of Rosenfeld & Raab (1974). Noteworthy also in the Algerian specimens of this collection is that many tend to become smooth in the anteromarginal region.

Genus *Monoceratina* Roth, 1928

Monoceratina? hodnaensis sp. nov.
Pl. 5:10–12

Synonymy. – □1985 *Pedicythere?* sp.2 – Vivière, p. 248, Pl. 25:15.

Holotype. – PMAL96 (Pl. 5:10). Bordj Ghdir section, level A13.

Derivation of the name. – After the provenance of the types in the mountain chain of Hodna, northeastern Algeria.

Material. – Half a dozen carapaces.

Occurrence. – Bordj Ghdir section, levels A13–A16.

Distribution. – Cenomanian of Algeria.

Measurements. – Length 0.65–0.67 mm, height 0.36–0.38 mm.

Diagnosis. – A species of *Monoceratina?* characterized by a smooth surface, a mamillate ventromedian spine, compressed anterior and posterior margins, and by lacking a dorsomedian sulcus.

Description. – The carapace is subrectangular in lateral aspect. Greatest height is at the mid-length. The valves are smooth, subequal and rather inflated, but compressed along the anterior and posterior margins. There is a well developed, ventromedian, mamillate spine and a distinct posterodorsal cauda, but no dorsomedian sulcus. The anterior is boldly and evenly rounded. The dorsal margin is straight. The compressed anterior half of the ventral margin slopes gently upwards towards the anteroventral angle;

posteriorly it ascends strongly towards the cauda. Internal features were not available for study.

Remarks. – What appear to be synonymous forms have been referred tentatively to *Pedicythere?* by Vivière (1985). However, apart from the fact that these specimens lack the characteristic dorsomedian sulcus, *Monoceratina* (or perhaps *Bythoceratina*) is considered a more appropriate determination.

Family Cytheruridae Müller, 1894

Subfamily Cytheropterinae Hanai, 1957

Genus *Metacytheropteron* Oertli, 1957

Metacytheropteron berbericus Bassoullet & Damotte
Pl. 6:1–2

Synonymy. – ☐1959 Ostracode M1 – Glintzboeckel & Magné, p. 64, Pl. 3:22. ☐1969 *Cytheropteron* M1 (Glintzboeckel & Magné) – Grekoff, Pl.1:3a–b. ☐1969 *Cytheropteron berbericus* n. sp. – Bassoullet & Damotte, p. 137, Pl. 2:7a–d. ☐1974 *Metacytheropteron berbericum* (Bassoullet & Damotte) – Rosenfeld & Raab, p. 12, Pl. 2:26–28; Pl. 5:3–4. ☐1975 *Metacytheropteron berbericus* (Bassoullet & Damotte) – Colin & El Dakkak, pp. 58–59, Pl. 2:8–11. ☐1978 *Metacytheropteron beberic us* (Bassoullet & Damotte) – Babinot, p. 21, Pl. 4:10. ☐1980 *Metacytheropteron berbericus* (Bassoullet & Damotte) – Ben Youssef, p. 89, Pl. 6:21. ☐1981a *Metacytheropteron berbericus* (Bassoullet & Damotte) – Bismuth *et al.*, pp. 225–226, Pl. 8:7–8. ☐1983 *Metacytheropteron berbericus* (Bassoullet & Damotte) – Gargouri-Razgallah, p. 150, Pl. 25:3–5. ☐1985 *Metacytheropteron berbericus* (Bassoullet & Damotte) – Vivière, pp. 251–252, Pl. 26:9–10.

Material. – A few carapaces.

Occurrence. – Bordj Ghdir section, level A18.

Distribution. – Cenomanian of Algeria, Tunisia (also in the Upper Albian), Egypt, Israel (also in the Lower Turonian) and Portugal. For references, see synonymy above.

Measurements. – Length 0.46–0.47 mm, height 0.26–0.27 mm.

Remarks. – This widespread species is the only one in the present collection previously reported from southern Europe (Babinot *et al.* 1978). Colin & El Dakkak (1975) noted some average-size differences among individuals of different populations. The latter authors also defined the finer details of the costal ornament; however, as far as can be judged from Colin & El Dakkak (1975), this arrangement and the outline tend to show some minor variation *viz a vis* the present individuals from Algeria and Israel.

Metacytheropteron sp. A
Pl. 6:3–4

Material. – Half a dozen right valves.

Occurrence. – Bordj Ghdir section, levels A1–A2, A4.

Measurements. – Length 0.51–0.56 mm, height 0.26–0.27 mm.

Description. – The dorsal and ventral margins are straight and converge towards the posterior, joining the anterior and posterior margins at more or less well expressed angles. The posterodorsal section is slightly concave and terminates in a caudal process. The anterior margin is broad and asymmetrically rounded. There is a weak eye-spot below the anterior cardinal angle. The lateral ornament consists of dense reticulations superimposed by two ridges; a median ridge extending over the major length of the carapace and a ventral ridge sweeping evenly parallel to the ventral margin. The ventral surface shows longitudinal ridglets. *Neocythere* sp. 220 in Rosenfeld & Raab (1984) is very close wih respect to lateral ornament.

Remarks. – The valves probably represent a new species. Internal features are too recrystallized to yield any information of significance. The assignation to *Metacytheropteron* is based mainly on the strong shape agreement with the type species *Metacytheropteron elegans* Oertli.

Metacytheropteron sp. B
Pl. 6:6–7

Material. – A few valves.

Occurrence. – Bordj Ghdir section, levels A2–A3.

Measurements. – Length 0.69–0.87 mm, height 0.46–0.48 mm.

Remarks. – These few valves also may represent a new species of *Metacytheropteron*. The lateral outline of the right valve almost coincides with that of *Metacytheropteron* sp. A., as does the costal arrangement, although the intercostal surface is smooth. The Iranian *Metacytheropteron* IRD2, in Grosdidier (1973), displays similar ornamentation. The left valve is higher and overreaches the dorsal margin of the right valve. Internal features are unknown.

Genus *Cytheropteron* Sars, 1866

Cytheropteron? sp.
Pl. 6:8–10

Material. – About 20 carapaces.

Occurrence. – Bordj Ghdir section, level A16.

Measurements. – Length 0.51–0.56 mm, height 0.15–0.20 mm.

Description. – This is a medium-sized, predominantly smooth form with a subelliptical lateral outline and a long upturned, caudal process. There is a distinct anterodorsal compression below the regularly arched dorsal margin, the latter being furnished with a rim, as is also (at least partly) the convex ventral margin. The ventral surfaces of the protruding, blade-like, ventrolateral alae are adorned with longitudinal ridgelets. A few 'ripples' are visible along the lateral edge of the anterior margin. The carapace is subrhomboidal in dorsal view. Internal features were not available for study.

Remarks. – This species is most similar to *Pedicythere?* sp. 1 (Vivière 1985), from which it is differentiated by its greater size, ventral surface ornament, and lack of spines posterior to the ventrolateral alae. The arched dorsal margin suggests that *Cytheropteron* is a more appropriate location for the forms described here than *Pedicythere*.

Genus *Eocytheropteron* Alexander, 1933

Eocytheropteron? sp.
Pl. 6:5

Material. – About 40 carapaces.

Occurrence. – Bordj Ghdir section, levels A11, A15, A16.

Measurements. – Length 0.33–0.36 mm, height 0.20–0.22 mm.

Remarks. – A number of small, subrhomboidal, cytheropteronoid specimens are tentatively placed here. The posterodorsal, posterior and ventral margins are nearly straight, whereas the margin extending from the greatest height to the anteroventral angle is curved, with a continuous shift in the radius of the curvature. The caudal process is relatively long, stout and located at mid-height. A major part of the lateral surface is coarsely perforated. There is a well defined ventrolateral swelling, a condition which has motivated the assignation to *Eocytheropteron* (cf. Alexander 1933, pp. 195–196). Internal features were not available for study.

Eocytheropteron glintzboeckeli Donze & Lefèvre
Pl. 6:11–13

Synonymy. – 1959 Ostracode K3 – Glintzboeckel & Magné, p. 62, Pl. 2:17. ☐1980 *Eocytheropteron glintzboeckeli* Donze & Lefèvre – Ben Youssef, p. 89, Pl. 4:21. ☐1981b *Eocytheropteron glintzboeckeli* n. sp. Donze & Lefèvre – Bismuth *et al.*, Pl. 1:13–16. ☐1981a *Eocytheropteron glintzboeckeli* Donze & Lefèvre – Bismuth *et al.* p. 226, Pl. 9:1–3. ☐1983 *Eocytheropteron glintzboeckeli* Donze & Lefèvre – Gargouri-Razgallah, p. 151, Pl. 27:6–8. ☐1985 *Eocytheropteron glintzboeckeli* Donze & Lefèvre – Vivière, p. 250, Pl. 26:6–7.

Material. – About twenty carapaces and valves.

Occurrence. – Bordj Ghdir section, levels A7, A10–A11, A13–A15. Also Djebel Semmama (Lower Cenomanian).

Distribution. – Upper Albian – Cenomanian of Algeria and Tunisia. For references see synonymy above.

Measurements. – Length 0.62–0.72 mm, height 0.28–0.33 mm.

Remarks. – The dorsal margin of this subovoid, alate species is evenly rounded in the left valve, but partly straight in the right valve, which is overlapped by the left. There is a short, rather pronounced caudal process, and a marginally rimmed, ventrolateral ala. The valve surface is predominantly smooth, although there is a median ridge which serves as an upper border for some crest-like irregularities on the surface of the ala. A posteroventral spine is visible in Pl. 6:11. The ventral surface is rhomboid and adorned with longitudinal ridgelets.

The hinge is typical for the genus and seems to support the present generic assignation, although the presence of alae may favour the allied genus *Cytheropteron* (Alexander 1933, p. 196).

Subfamily Cytherurinae Müller, 1894

Genus *Cytherura* Sars, 1866

'*Cytherura*' sp.
Pl. 7:1

Material. – A single carapace.

Occurrence. – Bordj Ghdir section, level A4.

Measurements. – Length 0.46 mm, height 0.23 mm.

Remarks. – This single, poorly preserved carapace is most closely comparable with specimens described by Vivière (1985) as *Cytherura? scabritia* [*nomen nudum*; cf. p. 3]. However, apart from ornamental differences, which tend to lack significance because of poor preservation, the present individual may be distinguished from *Cytherura? scabritia* by the more amplified ventral region, the more upturned posterior, and the pronounced eye-tubercle, all of which tend to contradict an assignation to *Cytherura* (as does its stratigraphical location).

The shape is subrectangular in lateral view; the entire surface is granular in its present state. The eye-tubercle is pronounced, the dorsal margin straight. Apart from a straight anterior part which joins the rounded anterior margin rather abruptly, the ventral margin is convex and compressed to form a keel-like structure as it sweeps towards the subdorsally caudate posterior.

Genus *Procytherura* Whatley, 1970

Procytherura? cuneata sp. nov.
Pl. 7:5–6

Synonymy. – ☐1985 *Procytherura?* sp. 2 – Vivière, pp. 244–245, Pl. 25:5.

Holotype. – PMAL115 (Pl. 7:5). Bordj Ghdir section, level A2.

Derivation of the name. – Latin *cuneatus*, wedge-shaped.

Material. – Half a dozen carapaces.

Occurrence. – Bordj Ghdir section, level A2.

Measurements. – Length 0.44–0.51 mm, height 0.20–0.26 mm.

Diagnosis. – A comparatively large, smooth and strongly wedge-shaped representative of *Procytherura?*, marked by a pointed posterior.

Description. – The valves are subequal, the surface smooth, and the outline wedge-shaped. The length varies markedly, but this is probably explicable in terms of sexual dimorphism. The anterior margin is broadly rounded, the dorsal and ventral margins weakly convex and converging strongly towards the pointed, almost caudate posterior. There is a significant lateral compression of the ventral margin from the mid-length posteriorly. Eye-spots seem absent. Internal features were not available for study and the generic assignation is therefore uncertain.

Remarks. – Despite conspicuous overlap along the dorsal margin and an Upper Cenomanian provenance, *Procytherura*? sp. 2 of Vivière (1985) is considered synonymous to the specimens described above. They also show close similarities to *Nigeroloxoconcha* gr. sp. 1 of Vivière (1985) which differs in having eye-spots and a rim along the entire circumference.

'Procytherura' sp.
Pl. 7:3–4

Material. – A dozen carapaces.

Occurrence. – Bordj Ghdir section, levels A1–A2.

Measurements. – Length 0.35–0.37 mm, height 0.17–0.19 mm.

Remarks. – The small specimens figured here have been labelled as a matter of convenience for want of a better evaluation. They are reminiscent of Gen. ind. DZ333 (Bismuth *et al.* 1981a), although their lateral outline is more rectangular and their eye-spots hardly visible.

The anterior is gently rounded, the posterior more narrowly so. The posteroventral region is marginally inclined and laterally compressed, being demarcated from the more inflated ventromedian valve region. The valves appear subequal. The surface is coarse and poorly preserved.

Genus *Eucytherura* Müller, 1894

Subgenus Vesticytherura Gründel, 1964

Eucytherura (*Vesticytherura*)? sp.
Pl. 7:2

Material. A single carapace.

Occurrence. – Bordj Ghdir section, level A2.

Measurements. – Length 0.46 mm, height 0.20 mm.

Description. – The small, compressed carapace, here doubtfully referred to *Eucytherura* (*Vesticytherura*), is elongate subrectangular in lateral aspect with nearly straight dorsal and ventral margins. The anterior margin is evenly rounded and more compressed than the main region of the carapace, as is also the well demarcated triangular posterior. The surface of the main region of the valves is smooth, though somewhat undulose in its peripheral parts due to the presence of weak tubercles. The eye-tubercle is prominent.

Eucytherura (*Vesticytherura*)? cf. *multituberculata* Gründel
Pl. 7:7–8

Synonymy. – See Weaver (1982, p. 88) for relevant details concerning the synonymy of *Eucytherura* (*Vesticytherura*) *multituberculata.*

Material. – A single carapace and a single valve.

Occurrence. – Bordj Ghdir section, level A11.

Distribution. – See Weaver (1982) for notes concerning the distribution of *Eucytherura* (*Vesticytherura*) *multituberculata.*

Measurements. – Length 0.35–0.36 mm, height 0.18–0.20 mm.

Description. – The carapace is subtriangular in lateral aspect, and compressed in dorsal and ventral views. The dorsal margin is straight, the ventral weakly sinuous. The left valve overreaches the right valve along the dorsal part of the almost evenly rounded anterior margin, which is armed with a few fairly pronouned denticles. The posterolateral region is 'stepped' and shows a narrow margin behind a lip-like swelling. The lateral surface is coarse and knotty. A few nodes or pore conuli are arranged in a V-formation over the median valve region. There are two lobate tubercles along the dorsal region and two more in the vicinity of the posteroventral angle. The eye-tubercle is prominent; a hinge-ear is absent. Of the internal features, only a recrystallized hinge is preserved. It appears to be of merodont type *sensu* Van Morkhoven (1962).

Remarks. – The external habitus coincides with certain species of *Eucytherura*, particularly *Eucytherura* (*Vesticytherura*) *multituberculata* as interpreted by Weaver (1982) and most closely the Upper Cenomanian individual figured from Buckland Newton (Weaver 1982, Pl. 16:23).

Family Trachyleberididae Sylvester-Bradley, 1948

Remarks. – The marginal areas, as referred to among the trachyleberidids described below, are defined as follows: The anterior (margin) runs from the hinge-ear and terminates anterior to the inflexure (if present) that denotes the beginning of the ventral (margin); the ventral (margin) usually prolongs continuously into the ventral section of the triangular posterior (margin); the latter section terminates at the posterior extremity; the dorsal section of the posterior (margin) is defined by the angulated junction with the dorsal margin (posterior cardinal angle) and the posterior extremity; the dorsal margin coincides with the hinge-line.

Subfamily Protocytherinae Ljubimova, 1955

Genus *Parexophthalmocythere* Oertli, 1959

Parexophthalmocythere sp.
Pl. 7:9–12

Synonymy. – ☐1959 Ostracode E8 – Glintzboeckel & Magné, p. 64. Pl. 3:32. ☐1980 *Cythereis* sp. – Ben Youssef, Pl. 4:17. ☐1981a Ostracode E8 Glintzboeckel & Magné – Bismuth *et al.*, p. 237. ☐1983 Ostracode E8 Glintzboeckel & Magné – Gargouri-Razgallah, p. 159, Pl. 32:3. ☐cf. 1985 *Parexophthalmocythere rhombusa* n. sp. – Vivière, pp. 170–171, Pl. 7:9–12 [unpublished name; *nomen nudum* herein (cf. p. 3)].

Material. – A dozen carapaces.

Occurrence. – Bordj Ghdir section, levels A7, A10, A13–A15.

Distribution. – Upper Albian and Cenomanian of Algeria and Tunisia. For references, see synonymy above.

Measurements. – Length 0.67–0.72 mm, height 0.26–0.28 mm.

Description. – The carapace is very compressed, particularly in the anterior and posterior regions. With straight dorsal and ventral margins that converge strongly towards the posterior, the carapace is wedge-shaped in lateral view. Greatest height is at the weakly or moderately developed left hinge-ear; there is no hinge-ear in the right valve. There is a distinct notch adjacent to the fairly pronounced eye-tubercle in the left valve. The latter valve overreaches the right along the dorsal section of the subacute posterior, less so along the broad and almost evenly rounded anterior. A stout ridge, which is polymorphic in that it may be either entirely smooth, or studded, or adorned with coarse denticles, runs along the anterior and posterior margins. The lateral surface is basically smooth and lacks a subcentral tubercle, but is adorned with two vigorous, stubby tubercles along the mid-dorsal, and along the mid-ventral region. A thin ridge-like connection is vaguely discernible between the two mid-ventral tubercles. Internal features were not available for study.

Remarks. – Vivière (1985) incorporated all forms in the above synonymy into *Parexophthalmocythere rhombusa* [*nomen nudum*], although the lateral outline of his specimens (Vivière 1985, Pl. 7:9–10) seems to differentiate them from the other forms as well as from my specimens.

The Bordj Ghdir section has yielded two specimens differing from those described (one is figured here) in that the carapace is slightly higher and posteriorly somewhat more truncated; moreover, there is a vaguely delineated subcentral tubercle, some small tubercles scattered on the lateral surface, and a flap-like rim along the hinge-ear of the left valve; furthermore, the posterior mid-dorsal tubercle is comparably longer and distally furcated.

It is uncertain whether these two specimens and those described above can be considered as distinct at the species level. Inasmuch as they occur together, the differences noted may be due to sexual dimorphism or ontogenetic factors.

Subfamily Trachyleberidinae Sylvester-Bradley, 1948

Genus *Pterygocythere* Hill, 1954

Pterygocythere sp.
Pl. 7:13–15

Material. – 20 complete carapaces, including juveniles.

Occurrence. – Bordj Ghdir section, level A10.

Measurements. – Length 0.99–1.02 mm, height 0.55–0.57 mm (adult specimens).

Description. – The carapace is large, smooth, and roughly subtrapezoidal in lateral view. The greatest height is at the weakly or modeately developed hinge-ear which is adorned with a well developed eye-tubercle (see dorsal view). The left valve overreaches the right along almost the entire circumference. The anterior margin is broad, oblique and feebly denticulate. There is a pronounced inflexure where the anterior margin joins the venter. The posterior, with its extremity at mid-height, displays a straight dorsal section, whereas its ventral section is convex and studded. The dorsal margin is straight and hidden posteriorly behind a ridge-like elevation. Except for being compressed anteriorly and posteriorly, the carapace is inflated and almost elliptical in dorsal and ventral views. The ventral valve surface is adorned with a few ridgelets.

The lateral surface is entirely smooth, lacking a subcentral tubercle. There is a straight, rather narrow, ventral carina which runs from a site slightly above the anteroventral angle and terminates abruptly near the posteroventral angle. Internal features were not available for study.

Remarks. – There are different opinions on whether *Pterygocythere* Hill, 1954, is to be considered a subgenus of *Pterygocythereis* Blake, 1933, or as a separate genus (cf. Babinot 1980; Van Morkhoven 1962; Weaver 1982; Reyment 1984). I have followed the steps taken by Babinot (1980) and Reyment (1984) in adopting the full generic differentiation of the two categories. However, irrespective of what is eventually decided, the present material agrees better with *Pterygocythere*, at least as far as can be judged from the external habitus. It is important to note that the type species of *Pterygocythereis*, i.e. *P. jonesi* (Baird), as well as several other representatives of that genus, are adorned with large spines, although less spiny representatives resembling the present specimens are also known (e.g. *P. miocenica* van den Bold).

Genus *Algeriana* gen. nov.

Type species. – *Algeriana reymenti* sp. nov.

Derivation of the name. – After the State of Algeria.

Diagnosis. – A new trachyleberidid genus with the following characteristics: Carapace large and heavily calcified; left valve larger than right; sexual dimorphism pronounced with lateral shape either subtrapezoidal or more elongate quadrangular; surface ornament features three lateral ridges consisting of partly or entirely disconnected, stubby tubercles; intercostal surface may yield a few additional tubercles but is predominantly smooth; subcentral tubercle feeble to moderate, though often cut up into pieces; hinge-ear feeble; eye-tubercle moderate and distinct; hinge holamphidont; central muscle-scars typically trachyleberidid comprising four adductors peculiarly arranged in a subvertical series plus a broadly V-shaped frontal scar.

Algeriana reymenti sp. nov.
Pl. 8:1–14

Holotype. – PMAL127a (Pl. 8:2). Bordj Ghdir section, level A12.

Derivation of the name. – After Professor R.A. Reyment, Uppsala.

Material. – Hundreds of carapaces and valves.

Occurrence. – Bordj Ghdir section, levels A10–A16.

Measurements. – Length males 1.13–1.23 mm, females 0.97–1.08 mm. Height males 0.59–0.62 mm, females 0.56–0.58 mm.

Diagnosis. – A species of *Algeriana* with a striate ventral surface.

Description. – The carapace is large, heavily calcified and compressed at both ends in dorsal view. Sexual dimorphism is pronounced; the forms interpreted as males are fewer, longer and more quadrangular in lateral view; the presumed females are shorter and subtrapezoidal in lateral aspect. The greatest height is at the feebly developed hinge-ear and equals about half the length in males, more than half the length in females. The left valve overreaches the right along almost the entire circumference, conspicuously along the dorsal section of the posterior margin and in front of the anterior cardinal angle. The eye-tubercle is moderately developed and distinct. The dorsal margin is straight. The anterior margin is broad, almost evenly rounded and generally adorned with coarse denticles; specimens with pronounced denticles tend to develop a groovy pattern along the edge of the anterior margin. The posterior margin is adorned with a few denticles along its convex ventral section; its dorsal section is straight. There is a marked inflexure where the anterior margin joins the convex ventral margin.

There are three lateral ridges consisting of half a dozen (usually one or two less on the median ridge) disconnected or partly fused, stubby or knobby tubercles. The intercostal surface is predominantly smooth but features a few additional tubercles, especially in front of the subcentral tubercle; the latter is feeble to moderate, often subdivided. The dorsal ridge is gently convex and overreaches the dorsal margin posteriorly; its two anterior tubercles are always disconnected, whereas the remaining tubercles are often partly or entirely fused. A similar arrangement concerns the ventral ridge which, however, displays three disconnected, posterior tubercles. There is a large bi- or tripartite boss obscuring the posteroventral angle below the ventral ridge. A few discrete tubercles mark the median ridge.

The ventral surface exposes a pair of straight ridgelets in each valve; the inner ridgelet prolongs the outer edge of the anterior margin and terminates slightly behind mid-length in females, whereas in males it joins up with a ventral protuberance of the posteroventral boss. The outer ridgelet is considerably shorter. The hinge is regarded as holamphidont, although the posterior tooth of the right hinge is poorly preserved.

Remarks. – Apart from *Algeriana cenomanica* (see below), *Cythereis aaramtaensis aaramtaensis* Bischoff, and *C. bleinensis sufflata* Donze, are probably the forms most closely comparable to the new genus with respect to external properties; their internal features, however, are typical for the genus *Cythereis*, according to Bischoff (1963) and Donze & Porthault (1972).

Algeriana cenomanica sp. nov.
Pl. 9:1–12

Synonymy. – ☐1959 Ostracode F3 – Glintzboeckel & Magné, p. 64, Pl. 3:30. ☐1980 *Cythereis aaramtaensis* cf. *aaramtaensis* Bischoff – Ben Youssef, p. 81, Pl. 6:12–13. ☐1981a *Cythereis* cf. *aaramtaensis aaramtaensis* Bischoff – Bismuth *et al.* p. 230, Pl. 9:13–15. ☐1983 *Cythereis* cf. *aaramtaensis aaramtaensis* Bischoff – Gargouri-Razgallah, p. 153, Pl. 28:6–7. ☐1985 *Cythereis* cf. *aaramtaensis aaramtaensis* Bischoff – Vivière, p. 173, Pl. 8:1–2.

Holotype. – PMAL141 (Pl. 9:5). Bordj Ghdir section, level A17.

Derivation of the name. – After the stratigraphical position of the species.

Material. – About twenty carapaces and a few single valves.

Occurrence. – Bordj Ghdir section, level A17. Also Djebel Semmama.

Distribution. – Cenomanian of Algeria and Tunisia. For references see synonymy above.

Measurements. – Length males 0.90–0.97 mm, females 0.77–0.79 mm. Height males 0.46–0.51 mm, females 0.41–0.49 mm.

Diagnosis. – A species of *Algeriana* that is smaller than the type species, rimmed along the anterior and posterior margin, and with a reticulate, ripple-like ventral surface.

Description. – This taxon agrees with the type species in overall shape, sexual dimorphism, carapace thickness, and size difference between the two valves. The broad anterior margin and the triangular posterior are strengthened by a rim adorned with denticles. The eye-tubercle is moderately large and distinct. The subcentral tubercle and the lateral ridges follow the arrangement of the type species, although the tubercles of the dorsal and the ventral ridge generally tend to be more sharply delimited from each other. The ventral ridge ascends slightly towards the posterior and ends in a large boss near the posteroventral angle. Compared with the type species, the intercostal surface may feature some additional tubercles. The ventral surface bears reticulations arranged in ripples.

The hinge appears to be holamphidont. The muscle-scars include a V-shaped frontal scar. The only clearly observable adductor muscle-scars are the lower ones of the left valve which naturally are reversed in relation to their counterparts of the right valve (cf. the type species).

Remarks. – This species is differentiated from *Algeriana reymenti* by its smaller size, antero- and posteromarginal rims, and ventral surface ornament. Contrary to previous provisional assignments (see synonymy above), it is relatively easy to distinguish between this species and *Cythereis aaramtaensis aaramtaensis* on the basis of the significant size difference and the internal features.

Stratigraphical note. – The two species of *Algeriana* have not been recorded together in the Bordj Ghdir section. There is a 6 m non-exposed sequence between the last appearance of *A. reymenti* and the single level with *A. cenomanica*. An evolutionary transition from the former to the latter, however, seems improbable on the limited data, especially taking into consideration the Bismuth *et al.* (1981a) and Vivière (1985) reports on Lower Cenomanian *A. cenomanica*.

Genus *Cythereis* Jones, 1849

'Cythereis' sp.
Pl. 10:1–4

Material. – 4 complete carapaces.

Occurrence. – Bordj Ghdir section, level A2.

Measurements. – Length males 0.82 mm, females 0.74 mm. Height males and females 0.41–0.43 mm.

Remarks. – These specimens are like *Algeriana* in possessing lateral ridges consisting of disconnected tubercles and a predominantly smooth intercostal shell surface. There is also a well developed posteroventral boss below the ventral ridge. The size and overall shape, however, differ markedly from *Algeriana*.

The carapace is subtrapezoidal to subtriangular in lateral aspect. Contrary to *Algeriana*, the hinge-ear is more pronounced as is also the eye-tubercle and particularly the subcentral tubercle. The anterior margin is rimmed and faintly studded, though not as broad and evenly rounded as in *Algeriana*. The triangular posterior margin is also rimmed and bears a few rounded denticles along its convex ventral section (see ventral view); its dorsal section is straight and reaches below mid-height. The ventral surface is coarse and irregular. Sexual dimorphism appears to be present, with the presumed males being longer.

Cythereis aff. *fahrioni* Bischoff
Pl. 10:5–12

Synonymy. – □aff.1963 *Cythereis fahrioni* n. sp. – Bischoff, p. 31, Pl. 12:90–93. □?1980 *Cythereis btaterensis btaterensis* Bischoff – Ben Youssef, Pl. 2:1–3.

Material. – A dozen carapaces.

Occurrence. – Bordj Ghdir section, levels A2, A4.

Measurements. – Length 0.74–0.87 mm, height 0.36–0.46 mm.

Description. – The carapace is subtrapezoidal in lateral aspect. The left valve overreaches the right along the dorsal section of the posterior margin and around the anterior cardinal angle. The hinge-ear is feebly to moderately developed. The eye-tubercle is prominent and surrounded by two notches. The dorsal margin is straight and converges towards the posterior with the ventral margin. The anterior margin is obliquely rounded and strengthened with a smooth rim which lacks denticles. The triangular posterior is compressed and marginally rimmed; its ventral section is curved while its dorsal part is straight and reaches below mid-height.

The lateral surface is mainly reticulate, displaying polygonal or rounded fossae; the posterolateral area is almost smooth as is also the anteromarginal region on some individuals; however, the latter region usually displays distinct, rectangular reticula. The subcentral tubercle is smooth and pronounced. There are three smooth and continuous lateral ridges. The dorsal ridge is gently convex and overreaches the dorsal margin, thus forming a shallow furrow along the valve border. It terminates posteriorly in a thin vertical ridgelet which connects transversely with the terminal endpoint of the thin, strutted, median ridge. The anteromarginal rim is similarly yoked to the ventral ridge; this latter terminates posteriorly in a distinct knob.

The carapace is ovate and compressed terminally in dorsal view. The ventral surface is regularly reticulate, displaying ovate fossae. There are some remarkable variations with respect to the outline of the ventral surface and also in overall size which are probably due to sexual dimorphism. Internal features were not available for study.

Generic assignment. – The lateral ridges and the general ornamental pattern are similar to species of the *Limburgina ornata* group (cf. Liebau 1971) from which the present forms may be distinguished by their more *Cythereis*-like triangular posterior.

The taxon qualifies better for admission into *Rehacythereis* Gründel. *Rehacythereis* was originally (Gründel 1973) distinguished from *Cythereis* by its continuous dorsal and ventral ridges and generally smooth subcentral tubercle. According to Gründel (1974), *Cythereis* is an Albian descendant of *Rehacythereis*, from which it has evolved towards its hallmark features, i.e disrupted dorsal and ventral ridges, and a generally tuberculated or reticulated subcentral tubercle. Babinot (1980, p. 123) considers this evolution (from entire to disrupted ridges) as possibly being a general trend among Albian–Cenomanian *Cythereis* forms.

Neale (1982) and Weaver (1982) adhered to Gründel (1973), with regard to the distinction between the two genera. However, Damotte (1977) critically appraised the independence of *Rehacythereis*, regarding it as poorly defined and difficult to maintain in view of some morphological intermediates. The situation is further confused by Gründel (1973) with his enigmatic assignment to *Rehacythereis* of certain species which clearly transcend the limits of that category (cf. Damotte 1977). Damotte (1977) finally reached a compromise and referred *Rehacythereis* and *Cythereis* (together with *Cornicythereis*) to different subgenera of the main genus *Cythereis*. In the present collection, however, there are obvious intermediates (between *Cythereis* and *Rehacythereis*). I have therefore suppressed *Rehacythereis* in order to avoid the arbitrary assignments of intermediates.

Remarks. – *Cythereis btaterensis btaterensis* Bischoff, as interpreted by Ben Youssef (1980, Pl. 2:1–3), probably coincides with the present material, but has merely a superficial resemblance to the original description of that species (Bischoff 1963). Of other species described by Bischoff (1963), the present specimens are most closely comparable with *Cythereis fahrioni* from which they are distinguished by straighter, better expressed lateral ridges, entirely smooth antero- and posteromarginal ridges, a subcentral tubercle and details of the ventral (in ventral view) and posterior outline.

Cythereis fahrioni bigrandis subsp. nov.
Pl. 11:1–13

Synonymy. – ☐ 1963 *Cythereis fahrioni* n. sp. – Bischoff, p. 31, Pl. 12:90–93; Pl. 13:94. ☐?1980 *Cythereis fahrioni* Bischoff – Ben Youssef, pp. 80–81, Pl. 4:1–5. ☐1981a *Cythereis* cf. *fahrioni* Bischoff – Bismuth *et al.* p. 231, Pl. 9:6–8. ☐1981a *Cythereis* aff. *fahrioni* Bischoff – Bismuth *et al.* p. 231, Pl. 10:8–11. ☐?1983 *Cythereis* aff. *fahrioni* Bischoff – Gargouri-Razgallah, p. 154, Pl. 30:1–7. ☐1985 *Cythereis* gr. *fahrioni* Bischoff – Vivière, pp. 172–173, Pl. 7:13–15.

Holotype. – PMTN24 (Pl. 11:1). Djebel Semmama.

Derivation of the name. – Latin *bi-*, two, and *grandis*, large, after its two co-existing size-classes.

Material. – About 30 carapaces and valves.

Occurrence. – Djebel Semmama (Lower Cenomanian).

Distribution. – Upper Albian – Cenomanian of Tunisia and Algeria (Albian of Lebanon). For references see synonymy above.

Diagnosis. – A polymorphic subspecies of *Cythereis fahrioni* which can be divided into two co-existing size-classes.

Measurements. – Larger forms (Tunisia): Length males 0.82–0.85 mm, length females 0.72–0.77 mm, height males 0.41 mm, females 0.41 mm. Smaller forms (Tunisia): Length 0.62–0.67 mm, height 0.31–0.33 mm. Bordj Ghdir specimens: Length male 0.92 mm, female 0.74 mm. Height male 0.46 mm, female 0.41 mm. Ras-El-Oued specimens: Length male 0.92 mm, female 0.77 mm. Height male and female 0.46 mm.

Description. – The carapace is medium-large to large, entirely reticulate and subrectangular to trapezoidal in lateral view. Sexual dimorphism is pronounced, with males being longer and more quadrangular. The greatest height is at the moderately strongly developed hinge-ear. The left valve overreaches the right along the dorsal section of the posterior margin. The eye-tubercle is hemispherical and prominent. The anterior margin is broadly, though somewhat obliquely rounded, and adorned with coarse denticles alternating with minute papillae along its front edge (see ventral view). As seen in lateral view, there is a thin rim which prolongs the eye-tubercle and runs parallel to the anterior margin (sometimes coinciding with it ventrally) all the way down to the junction with the ventral margin. That rim is adorned with short pore conuli or sieve plates which ventrally are sometimes celate. The ventral margin appears sinuous due to a lengthy inflexure at the junction with the anterior margin. The dorsal margin is straight. The posterior margin is triangular in the right valve, more rounded in the left, and, at least along its convex ventral section, is adorned similarly to the anterior margin (see ventral view).

The lateral surface is regularly reticulate, consisting of polygonal fossae with smooth sola. Single, intra-mural pores are present at the junction of some muri. The only discernible disruptions of the reticulate framework are near the subcentral tubercle and posteroventrally. As can be seen on tilted specimens, there are plications on the valve surface, producing a subcentral tubercle and three lateral ridges (dorsal, median and ventral) which are hardly discernible in lateral view, being camouflaged by the reticulum, although a short flexuous ridge is visible on the median surface. The dorsal ridge overreaches the dorsal margin and forms a shallow furrow along the valve border (see dorsal view). The ventral surface is triangular and covered with rounded fossae. Greatest width lies across the posteromedian region.

The hinge is paramphidont, in that the elements of the right valve consist of an anterior lobed boss with three bulla-like crenulae, an anteromedian socket bounded ventrally by a concave ridgelet, a smooth posteromedian furrow, and an elongate, crenulate posterior tooth. The marginal pores, which reach into the marginal denticles, are straight, single and fairly numerous. The inner lamellae are moderately broad. The muscle-scar pattern consists of a V-shaped frontal scar, and four rod-like adductor scars in a subvertical series.

The internal as well as the external features are typical of *Cythereis, sensu* Van Morkhoven (1963, pp. 179–182) and Damotte (1977).

Remarks. – Although a few conspecific individuals have been recorded from the Bordj Ghdir section, the description above is based on the more numerous and better preserved specimens from Tunisia. These forms are particularly interesting in that they can be divided into two co-existing size-classes (i.e. they are size polymorphic); sexual dimorphism is pronounced in the larger size-class, but is not readily observable in the smaller group. Being externally as well as internally identical, there is no doubt that all these specimens belong to the same species, although some smaller representatives tend to develop a more pointed posterior.

Hartmann (1982) recorded similar size-differences among individuals within populations of *Xestoleberis chilensis austrocontinentalis* Hartmann, and suggested that this might be explained by variations in the ecological conditions under which the different generations grew.

The Algerian specimens (Pl. 12:1–2; i.e. the two individuals recorded from the Bordj Ghdir section, level A15), which are here considered as conspecific with the Tunisian forms, differ somewhat from the latter in that the ventral ridge is partly or entirely disrupted into disconnected bullae; moreover, the postero-terminal knob of this ridge appears comparably more pronounced. The male individual also displays some sparsely scattered papillae over the lateral surface.

Two additional specimens (Pl. 12:3–4), recorded from the Ras-El-Oued area and of equivalent age to those of the Bordj Ghdir section, are figured here. They seem to coincide with those described, although all hard cement from the carapace surface could not be removed, thus preventing comparisons of ornamental details. However, the posterior margin of the male representatives are somewhat more truncated compared with those described above.

Assignment. – There are distinct ornamental differences between the specimens treated here and those referred to *Cythereis* aff. *fahrioni* earlier, differences which seem to support a distinction at least at the species level. However, it is very difficult to say which of these two taxa show the strongest affinity with *Cythereis fahrioni*, although the dissimilarities between the latter (as described by Bischoff 1963) and the forms described here, are considered subspecific. The subspecies is mainly differentiated by having two co-existing size-classes, possibly also by less distinct lateral ridges and details of the anteromarginal ornamentation.

Cythereis namousensis Bassoullet & Damotte
Pl. 10:13–16

Synonymy. – ☐1969 *Cythereis namousensis* n. sp. – Bassoullet & Damotte, pp. 134–135, Pl. 1:3a–d. ☐1974 *Cythereis namousensis* Bassoullet & Damotte – Rosenfeld & Raab, p. 17, Pl. 3:17–18. ☐1980 *Cythereis namousensis* Bassoullet & Damotte – Ben Youssef, pp. 78–79, Pl. 6:14–16. ☐1981a *Cythereis namousensis* Bassoullet & Damotte – Bismuth *et al.*, p. 232, Pl. 9:9–10. ☐1983 *Cythereis namousensis* Bassoullet & Damotte – Gargouri-Razgallah, p. 154, Pl. 29:1. ☐1985 *Cythereis namousensis* Bassoullet & Damotte – Vivière, pp. 174–175, Pl. 3:6–7.

Material. – About 30 juveniles.

Occurrence. – Bordj Ghdir section, level A16.

Distribution. – Cenomanian of Algeria, Tunisia, Israel. For references see synonymy above.

Remarks. – About 30 immature individuals of what seems to be *Cythereis namousensis* are briefly considered here. Apart from their smaller size, they differ from the adults mainly in that the subcentral tubercle is poorly developed and the ventral ridge not clearly delimited from the surrounding reticula. Moreover, the two separate groups of tubercles which denote the main components of the 'dorsal ridge' (Bassoullet & Damotte 1969, pp. 134–135), are strongly reduced each being merely a node-like elevation.

Some Israeli specimens interpreted as *C. namousensis* by Rosenfeld & Raab (1974) are figured here (Pl. 10:15–16) for comparison. Apart from being slightly larger than the material of Bassoullet & Damotte (1969), they seem to be genuine *C. namousensis*. Although the tubercles of the 'dorsal ridge' are almost as reduced as those of the instars described above, their overall shape and other features of the surface ornament strongly tie these Israeli representatives to the association originally described by Bassoullet & Damotte (1969).

Cythereis? punctatafoveolata sp. nov.
Pl. 12:5–13

Holotype. – PMAL161 (Pl. 12:6). Bordj Ghdir section, level A15.

Derivation of the name. – Latin *punctatus* and *foveolatus*, after the centrally punctate and marginally foveolate lateral surface.

Material. – About two dozen carapaces and valves.

Occurrence. – Bordj Ghdir section, levels A10, A12, A15.

Measurements. – Length males 0.69–0.74 mm, females 0.64–0.69 mm. Height males 0.36–0.38 mm, females 0.37–0.38 mm.

Diagnosis. – Carapace of moderate size; sexual dimorphism pronounced; lateral surface centrally punctate, marginally foveolate; lateral ridges thin and continuous; hinge-ear moderately developd; eye-tubercles prominent; anterior hinge-element of the right valve smooth and kidney-shaped.

Description. – The sexual dimorphism is pronounced, with males being longer, less inflated and more rectangular compared with the subtriangular females. The left valve is the larger, overreaching almost the entire circumference of the right. The hinge-ear is feebly developed. The eye-tubercle is prominent and borders an adjacent notch in the left valve. The dorsal margin is straight and partly obscured by the dorsal ridge. The anterior margin is about evenly rounded and joins the venter at an inflexure; it is armed with a weak rim which tends to be more developed in the left valve. At least ventrally (see ventral view), this rim is adorned with pronounced lobe-like denticles which become smaller or vanish entirely towards the dorsum. The posterior is triangular in the right valve, more rounded in the left. The shape difference between the dorsal and ventral sections of the posterior, is only distinct in the right valve. The posteromarginal rim, which is denticulate like the anterior margin, is only distinct in the right valve.

The central region of the lateral surface is punctate, whereas the antero- and posteromedian regions are punctate–foveolate; the marginal regions display sparsely distributed foveolae. The subcentral tubercle is moderately developed and pitted. The ventromedian region displays a single spine or pore-conulus. There are three thin and continuous lateral ridges. The median ridge, which is disconnected from the subcentral tubercle, is strutted with half a dozen nodes. The dorsal ridge is slightly arched and displays two spiny denticles anteriorly. As seen in dorsal view, the two dorsal ridges run parallel on females, whereas they diverge distinctly towards the posterior on males. The ventral ridge is disconnected from the anteromarginal rim and terminates posteriorly with a prominent knob.

The carapace is ovate to hexagonal in dorsal and ventral views. The triangular ventral surface is reticulate, showing longitudinal ridges interconnected by transverse ridgelets; the fossae are rectangular and usually contain secondary pits.

The hinge is amphidont. The anterior tooth of the right hinge is smooth and kidney-shaped; the corresponding socket of the left valve is open ventrally and located below a swollen overhang.

Remarks. – The shape of the smooth anterior tooth of the right valve hinge differs markedly from the crenulated morphology usually associated with representatives of *Cythereis*. Apart from the hinge, however, internal features could not be studied to confirm such further differences as might justify erection of a new genus, although the new species is rather easily differentiated from other species of *Cythereis* or any other known trachyleberidid genus by details of the surface ornament.

Cythereis afroreticulata sp. nov.
Pl. 13:1–8

Holotype. – PMAL169 (Pl. 13:1). Bordj Ghdir section, level A14.

Derivation of the name. – After its provenance and ornament.

Material. – Hundreds of carapaces and valves.

Occurrence. – Bordj Ghdir section, levels A10–A16.

Measurements. – Length 1.08–1.28 mm, height 0.59–0.64 mm.

Diagnosis. – A species of *Cythereis* with the following characteristics: Large, trapezoidal to subtriangular carapace with a reticulate surface; three lateral ridges consisting of disconnected tubercles; ventral ridge separate from the anteromarginal rim; ventral surface with shallow anterior furrows; anterior tooth of the right hinge smooth.

Description. – The carapace is large, heavily calcified and displays a shape typical for the genus. The left valve overreaches the right along the dorsal section of the posterior margin and around the anterior cardinal angle. The hinge-ear is moderately or feebly developed in the left valve,

absent in the right valve. There is a large notch in front of the prominent eye-tubercle in the left valve. The dorsal margin is straight and the ventral margin slightly convex, partly due to a slight inflexure at the junction with the anterior. The broad anterior margin is evenly rounded and rimmed. The flat lateral edge of this rim is sparsely noded; however, as seen in dorsal view, there are two parallel rows of minute denticles running along the anterior front; those of the outer row gradually develop into coarse denticles ventrally (see ventral view). The triangular posterior is irregularly rimmed and subacute below mid-height; its dorsal section is straight in the left valve and slightly concave in the right; its ventral section is convex.

The lateral surface is reticulate with rather deep fossae and stout muri. The fossae of the median and posteromedian region are polygonal; those of the anteromedian region are rounded. The anterior region exhibits half a dozen muri which join the anteromarginal rim at regular intervals. The subcentral tubercle is moderately sized and irregularly ornamented. The median ridge consists of half a dozen disconnected nodes (some of which are hollow) positioned at the junction of muri. A few stout, short and disconnected tubercles overreach the dorsal margin and constitute the dorsal ridge. The ventral ridge consists of half a dozen separated, prominent tubercles, the posteriormost being particularly powerful.

The carapace is subhexagonal in dorsal view. The ventral surface is reticulate, and displays an anterior furrow crossed by transverse filaments in each valve; these furrows are shallower than the surrounding fossae. The hinge of the right valve, which is hemiamphidont, displays a smooth, rounded anterior tooth, an ovate anteromedian socket, and a thin posteromedian furrow followed by an elongated posterior boss. The marginal pore canals are sparse and straight. The zone of concrescence is broad anteriorly and posteriorly.

Juveniles are abundant showing only rounded reticula in the median region and less expressed median and ventral ridges.

Generic remarks. – The anterior hinge-tooth of the right valve is smooth, but not as extraordinarily shaped as in *C.? punctatafoveolata*. In this and its external aspects, at least compared with *C.? punctatafoveolata*, the species falls more conveniently within the permitted intra-generic variation of *Cythereis* (cf. Van Morkhoven 1962, p. 80).

Remarks. – Reticulate forms similar to this species are rather common and widespread in the European realm during the Middle and Upper Cretaceous. Forms of Albian–Cenomanian age have often, albeit sometimes arbitrarily, been referred to *Cythereis reticulata* (Jones & Hinde), whereas younger forms often have been labelled *Cythereis ornatissima* (Jones & Hinde) (cf. Kaye 1964). There are several incompatible concepts of *C. reticulata* in the literature. I will not attempt the needed revision herein, but only pin-point certain morphological features displayed by Albian–Cenomanian *reticulata*-forms similar to the new species, features which can be used to distinguish them from the new species.

Triebel (1940) assigned forms to *C. reticulata* which display a crenulated anterior right-hinge tooth as well as entire dorsal and ventral ridges which are not developed into disconnected tubercles. In specimens referred to the species by Kaye (1964), Damotte & Grosdidier (1963), and also in some representatives of *C. reticulata* in Deroo (1956) and in *C. hirsuta* Damotte & Grosdidier, the overall shape is more rectangular in lateral view and the anteromarginal rim is prolonged into the ventral ridge; there are also prominent antero- and postermarginal spines. The specimens referred as *C.* aff. *reticulata* by Donze & Thomel (1972), and those illustrated by Weaver (1982), also show a continuity between the anteromarginal rim and the ventral ridge, the latter being entire and strutted. Further, forms assigned to *C. reticulata* by Donze & Porthault (1972) display a very well defined posterior region and an alate ventral ridge which prolongs the anteromarginal rim.

Some authors, however (e.g. Donze 1972), seek to explain many of these morphological differences in terms of sensitivity to ecological changes, i.e. ecophenotypy.

There are only a few comparable forms from the Mesogean part of the Tethys sea. The most similar is probably *Cythereis algeriana* Bassoullet & Damotte, as interpreted by Rosenfeld & Raab (1974), which is differentiated by its entire ventral ridge and details of the ventral surface ornament.

Genus *Veeniacythereis* Gründel, 1973

Veeniacythereis? spp.
Pl. 13:9–11

Material. – A single carapace from the Bordj Ghdir section and a few carapaces of somewhat uncertain provenance.

Occurrence. – Bordj Ghdir section, level A7.

Measurements. – Length 0.87 mm, height 0.41 mm.

Description. – The carapace is strongly calcified and subrectangular to subtriangular in lateral aspect. The entirely smooth lateral surface is corrugated by its ornament of three bulky ridges. The left valve is the larger, overreaching the right along the dorsal section of the symmetrically triangular posterior around the hinge-ear. The hinge-ear is fairly prominent in the left valve, less developed in the right. Eye-tubercles are not well expressed. The anterior margin is almost evenly rounded in the right valve, more obliquely so in the left valve. There is a thick anteromarginal rim which lacks denticles but prolongs the swollen ventral ridge; the latter thickens progressively towards the posterior. The posterior is subacutely pointed at midheight and rimmed, at least along its straight ventral section where a few studs are discernible. The dorsal margin is straight and depressed in a furrow formed by the overhang of the dorsal ridge. The dorsal ridge, which is more pronounced in the left valve, terminates posteriorly in a downward directed deflection. The bulky median ridge is not sharply delineated but accommodates a feebly developed subcentral tubercle. The carapace is subelliptical in dorsal view. The furrow walls of the dorsal ridges show a few

transverse ridgelets; the triangular ventral surface bears straight ridgelets.

A few additional carapaces of uncertain provenance (in the Ras-El-Oued region) are also referred here. Their general habitus strongly resembles the one described (from the Bordj Ghdir section), although they are differentiated by a shorter median ridge, more sharply delineated anteromarginal and dorsal ridges, and by the presence of pits in the ventromedian region. *Cythereis?* sp. of Vivière (1985 Pl. 8:3–5) is probably conspecific with these specimens.

Remarks. – Possibly half a dozen genera include representatives possessing a lateral frame reminiscent of the individuals discussed above (cf. Reyment 1984, pp. 88–91). The present specimens are, however, referred to *Veeniacythereis* by virtue of their posterior development. Although the gross morphology of the lateral ridges may attain great differences among specimens referred to that genus (cf. Damotte 1977; Gündel 1973), the present individual agrees comparatively well in this respect with the type species *Veeniacythereis imparia* Gründel, which, however, has a more boldly asserted eye-tubercle.

'Veeniacythereis' subrectangulata Majoran
Pl. 13:12–16

Synonymy. – □ 1988 *'Veeniacythereis' subrectangulata* sp. nov. – Majoran, pp. 696–697, Pls. 8, 9.

Material. – About a hundred carapaces and valves.

Occurrence. – Bordj Ghdir section, levels A10, A12–A15.

Remarks. – This species and its affinity to, e.g., *'Veeniacythereis' jezzineensis* Bischoff, and *'Veeniacythereis'* sp. nov?, have been treated in detail in Majoran (1988). To that discussion, I would like to add that *'Veeniacythereis' subrectangulata* and *'Veeniacythereis'* sp. nov? in the present collection, are each represented by only a single ontogenetic stage, a phenomenon possibly explained in terms of an allochthonous origin.

A single specimen of *'Veeniacythereis'* sp. nov? was recorded from the Bordj Ghdir section, but there are several from Djebel Semmama (Pl. 13:17). A few specimens were also obtained from level A16 (Bordj Ghdir section) and labelled *'Veeniacythereis'* aff. *streblolophata* (Al-Abdul-Razzaq) in Majoran (1988) though not figured here.

Genus *Maghrebeis* Majoran, 1987

Maghrebeis tuberculata Majoran
Pl. 14:1–3

Synonymy. – □ ?1985 *Eucythere? totaliana* n. sp. – Vivière, p. 245, Pl. 25:8–11 [unpublished name; *nomen nudum* herein (cf. p. 3)]. □ 1987 *Maghrebeis tuberculata* gen. et sp. nov. – Majoran, pp. 29–32.

Material. – About a hundred carapaces and valves.

Occurrence. – Bordj Ghdir section, levels A10–A14, A16.

Remarks. – This species was described in detail in Majoran (1987a). Since the appearance of this article, I have seen Vivière's unpublished thesis (Vivière 1985), and should like now to suggest the possibility that *Eucythere? totaliana* Vivière [*nomen nudum*], may be conspecific with, or closely related to, *Maghrebeis tuberculata*, although it is significantly smaller and has a more obliquely rounded anterior (Vivière 1985, pp. 245–247).

Genus *Peloriops* Al-Abdul-Razzaq, 1979

Peloriops ziregensis? (Bassoullet & Damotte)
Pls. 14:4–13; 15:1–3

Synonymy. – ☐1959 Ostracode E3 – Glintzboeckel & Magné, p. 62, Pl. 2:20. ☐?1969 *Cythereis ziregensis* n. sp. – Bassoullet & Damotte, p. 135, Pl. 1:4a–d. ☐?1974 *Planileberis ziregensis* (Bassoullet & Damotte) – Rosenfeld & Raab, p. 19, Pl. 3:1. ☐1979 *Peloriops sphaerommata* n. gen. et n. sp. – Al-Abdul-Razzaq, pp. 47–50, Pl. 1:1–3. ☐1980 *Cythereis ziregensis* Bassoullet & Damotte – Ben Youssef, p. 80, Pl. 6:17–18. ☐1981a *Peloriops ziregensis* (Bassoullet & Damotte) – Bismuth *et al.*, p. 234, Pl. 8:9–12. ☐1983 *Peloriops sphaerommata* Al-Abdul-Razzaq – Gargouri-Razgallah, p. 156, Pl. 31:1–2. ☐1985 *Peloriops ziregensis* (Bassoullet & Damotte) – Vivière, p. 201, Pl. 14:2–3.

Material. – About 50 carapaces and valves.

Occurrence. – Bordj Ghdir section, levels A10, A12–A16. Also Djebel Semmama (Lower Cenomanian).

Distribution. – See remarks below.

Measurements. – Length males 0.82–0.92 mm, females 0.67–0.77 mm. Height males 0.41–0.44 mm, females 0.41 mm.

Description. – The carapace is subrectangular to wedge-shaped in lateral aspect. Sexual dimorphism is pronounced; individuals interpreted as females are comparatively shorter than the presumed males and display a stronger convergence towards the posterior between the dorsal and ventral margins. There is a slight inflexure where the almost evenly rounded anterior margin joins the ventral margin; the latter becomes progressively more convex towards the posterior and merges into the arched ventral section of the subacute posterior. The left valve overreaches the right conspicuously along the concave dorsal section of the posterior margin. The dorsal margin is feebly convex. As seen in dorsal view, the prominent hinge-ear (of the left valve) displays a kidney-shaped protuberance which overlaps the right valve just behind the 'hinge-ear'. The eye-tubercle is very prominent, smooth and spherical or semi-ovate.

The lateral surface lacks ridges as well as a subcentral tubercle, but is entirely and closely foveolate (i.e. covered with minute pits), a pattern which is randomly knitted by tiny filaments; the latter tending to delimit subgroups of foveolae. Papillae (sometimes developed as pore-conuli) may be numerous; they are generally located at the junctions of filaments in the median valve region. There is a well developed pore-conulus situated half way between the

eye-tubercle and the large posterodorsal tubercle. The latter displays a crater-like top containing minute pits. There is usually a thin, vertical, isolated bulla postjacent to the posterodorsal tubercle, another bulla occurs on the corresponding side of the posteroventral tubercle. The foveolae of the anterolateral area are less distinct, although about five tiny filaments may be revealed perpendicularly joining a thin anteromarginal rim at regular intervals. That rim forms a postjacent border towards a belt of minute pits wich runs along the outermost lateral edge. A similar belt, delimited by a thin rim, is found on the outermost posterolateral edge. It has variably developed foveolae. The anterior margin and the posterior are usually adorned with more or less pronounced spiny denticles.

In dorsal and ventral view, the carapace is subhastate. The dorsal surface consists of foveolae superimposed by tiny filaments and peripheral papillae. The ventral surface is also foveolate and covered with straight ridgelets.

The hinge is amphidont. The anterior tooth of the right valve is irregularly notched along its sides; the anteromedian socket of the same valve is open ventrally; the posteromedian furrow is feebly convex; the posterior tooth is smooth with a subtriangular and flat top.

Comparison. – The lateral surface described above, with its dense network of foveolae superimposed by tiny filaments, is reminiscent of certain representatives of *Mauritsina* (cf. Reyment 1984). Moreover, apart from lacking foveolae, similar filamentous structures uniting papillae, are also known to occur in *Echinocythereis*, a condition penetrated in further detail by Reyment (1985).

Remarks. – Similar or identical forms to those discussed above have been encountered frequently in Cenomanian deposits along the southern border of the Tethys sea, ranging geographically from Algeria in the West to Israel and Kuwait in the East (cf. references in the synonymy above).

Apart from Algerian (Ras-El-Oued) representatives, the description is also based on identical forms from Tunisia (Djebel Semmama), figured previously by Bismuth *et al.* (1981a).

Cythereis ziregensis, as originally described and figured by Bassoullet & Damotte (1969) from the Upper Cenomanian of West Algeria, is smaller, ranging in length from 0.55–0.58 mm and in height from 0.30–0.31 mm. Further differences occur in the hinge as described by Bassoullet & Damotte (1969). However, Al-Abdul-Razzaq (1979) reported the hinge of *ziregensis* to be typical for the genus *Peloriops*, and its general habitus as closely resembling the type species, i.e. *Peloriops sphaerommata*. Indeed, later (Al-Abdul-Razzaq 1983) referred to *ziregensis* and *sphaerommata* as synonyms. *Peloriops sphaerommata* agrees in size with *ziregensis*, but otherwise seems more like the larger forms described above (cf Al-Abdul-Razzaq 1979).

As far as the present Algerian material is concerned, one of the sampling levels of the Bordj Ghdir section and one additional sample of somewhat uncertain provenance, have yielded a few specimens of small size, ranging in length from 0.36 to 0.45 mm and in height from 0.22 to

0.25 mm. The smaller individuals of the Bordj Ghdir section (Pl. 15:1–2) differ from their larger companions only in that the posterodorsal and the posteroventral tubercle tend to fuse more strongly with the corresponding postjacent bullae while the posterior margin of the left valve lacks coarse denticles. Apart from the size difference, these differences are minor, and may be regarded as intraspecific, possibly the outcome of interaction with ecological factors.

Further differences, which may possibly transcend the species level, characterize the small Algerian individuals of uncertain provenance (Pl. 15:3). These specimens display a tendency towards smoothening of the posteromedian region between the posterodorsal and the posteroventral tubercles which are smooth and exaggerated, much exceeding the eye-tubercle in size. The anterior and the posterior margins of these specimens are very swollen and lack spines or denticles.

Peloriops pustulata (Rosenfeld), erected by Rosenfeld & Raab (1974), is closely comparable with the present material. For example, the dorsal arrangement of papillae (cf. Rosenfeld & Raab 1974, Pl. 61) is almost identical with the large forms above, in this respect better resembling them than the Israeli fossils referred by them to *ziregensis* (Rosenfeld & Raab 1974).

Some Israeli forms are figured for comparison (Pl. 15:6–9), but they are too poorly preserved to allow proper identification. Nevertheless, in dorsal view they may be distinguished from the Algerian and Tunisian material in that they display a rim-like elevation along the dorsal margin which encloses a shallow furrow between the two valves.

Peloriops sp. nov?
Pl. 15:4–5

Synonymy. – ☐?1985 *Peloriops* cf. *elassodictyota* Al-Abdul-Razzaq – Vivière, p. 201, Pl. 14:4–7.

Material. – 5 carapaces.

Occurrence. – Bordj Ghdir section, level A10, A12.

Measurements. – Length 0.67–0.68 mm, height 0.37–0.38 mm.

Remarks. – These specimens can be differentiated from *Peloriops ziregensis* by the shape and perforate design of the single posterodorsal and posteroventral tubercles present, the broader and entirely pitted anteromarginal rim, the truncated and smooth posterior, the more numerous lateral papillae, and by the eye-tubercle being partly perforate.

Genus *Matronella* Damotte, 1974

Matronella sp.
Pl. 15:10–11

Material. – A single carapace.

Occurrence. – Bordj Ghdir section, level A1.

Measurements. – Length 0.90 mm, height 0.41 mm.

Description. – With almost straight dorsal and ventral margins conspicuously converging towards the posterior, the present individual has a subtriangular to trapezoidal shape in lateral view. The eye-tubercle is fairly prominent as is also the hinge-ear of the left valve. The anterior margin is broad, almost evenly rounded, and strengthened with a prominent ridge which is composed of closely arranged, more or less lobe-like elements. The anterior rim is prolonged into the ventral ridge which consists of a few, small, disconnected tubercles and a well developed, posteroterminal, multifurcated protuberance. The subcentral tubercle is large and adorned with a few secondary denticles. The median ridge consists merely of two disconnected tubercles. The dorsal ridge, with its irregular arrangement of small unlinked tubercles, terminates posteriorly in a strong, irregularly multifurcate or castellated process. The surface between these lateral ridges is smooth. The triangular ventral surface is irregular.

Remarks. – The external habitus of this carapace strongly resembles that of *Matronella*, particularly the type species *M. matronae* (Damotte & Grosdidier). However, neither the muscle-scars, which serve as the main hallmark of the genus (cf. Damotte 1974), nor any other internal features, could be studied. The type species differs somewhat from the present specimen in having spinier ridge-elements, and also in general shape (cf. Damotte 1974).

Genus *Curfsina* Deroo, 1966

Curfsina? sp.
Pl. 15:12–13

Material. – 2 complete carapaces and one damaged carapace.

Occurrence. – Bordj Ghdir section, levels A4–A5.

Measurements. – Length 0.82–0.87 mm, height 0.41 mm.

Description. – The carapace is robust, trapezoidal in lateral view, hexagonal in dorsal view. The valves are subequal, the entire surface being covered with a dense network of rounded foveolae. The lateral surface is weakly plicated by three prominent ridges. The hinge-ear is moderately pronounced; eye-tubercles are absent. The dorsal margin is straight, subparallel with the venter, and hidden behind the slightly arched dorsal ridge. The anterior margin, which is swollen but not rimmed, is obliquely rounded because of a straighter dorsal section. Similarly, the more compressed posterior, displays a long, straight dorsal section, whereas its ventral section exposes a sharp curvature below the level of the ventral margin. The median ridge is broad, flat and only vaguely disconnected from the weakly developed subcentral tubercle. The thick ventral ridge, which fuses anteriorly into the swollen anterior margin, becomes progressively thicker towards the posterior. The hastate ventral surface shows a few longitudinal ridgelets.

A second complete specimen differs somewhat from the one described above in that the carapace is slightly longer, less inflated, and displays a few posteroventral spines. The anterior margin is evenly rounded, less swollen and adorned with thorny denticles along its entire length.

Remarks. – These specimens are doubtfully referred to *Curfsina*. Contrary to what pertains in our material, representatives of that genus usually have eye-tubercles and an anteromarginal ridge (Deroo 1966, p. 139).

Gründel (1973) and Damotte & Rey (1980), however, wish to keep such forms within *Rehacythereis*; for example, the Barremian *Cythereis (Rehacythereis) ilhaensis* Damotte, which resembles the present specimens ornamentally, but differs in that the posterior ridge, the median ridge and the hinge-ear are less pronounced. The Aptian *Cythereis geometrica* Damotte & Grosdidier, and *C. semiaperta* Damotte & Grosdidier, which were referred to *Rehacythereis* by Gründel (1973), may easily be distinguished from the present specimens by the presence of an anteromarginal ridge, the feebly developed median ridge, and by the pronounced posterior termination of the dorsal ridge.

Genus *Trachyleberidea* Bowen, 1953

Trachyleberidea? sp.
Pl. 15:14–16

Material. – 4 carapaces.

Occurrence. – Bordj Ghdir section, levels A8–A9.

Measurements. – Length male 0.64 mm, females 0.56–0.59 mm. Height male and females 0.30–0.31 mm.

Description. – Four variably preserved specimens, which seem to belong to a single species, are considered here. Their minor differences seem sexually dimorphic, i.e. one of the specimens, here interpreted as a male, is somewhat longer and more rectangular in lateral view than the others.

The carapace is small, compressed and subrectangular to subtriangular in lateral aspect. The surface is covered with more or less closely dispersed foveolae. A faintly studded rim runs along the somewhat obliquely rounded anterior margin. This rim is weakly linked to the ventral ridge which terminates posteriorly in a minor swelling. The moderately developed dorsal ridge displays a somewhat 'chipped' outline, and terminates posteriorly in a vertical protuberance. There is no median ridge; a knob-like tubercle occupies this site. The subcentral tubercle comprises a weak, elongate elevation of the valve surface. Eye-tubercles are not discernible on the crescent-shaped ridge along the weakly developed hinge-ear. The posterior margin has a convex ventral section and a straight dorsal part which extends below mid-height.

Remarks. – As far as can be judged from the external habitus, the present specimens seem to qualify for admission into *Trachyleberidea*, although lacking the sharply pointed posterior usually associated with representatives of that genus (Babinot 1980, pp. 153–160). They resemble *Cythereis zoumoffeni zoumoffeni* Bischoff, from which they can be differentiated by the absence of eye-tubercles and posteroventral denticles.

Genus *Platycythereis* Triebel, 1940

Platycythereis? sp. juv.
Pl. 16:1

Material. – About 20 valves.

Occurrence. – Bordj Ghdir section, level A11, A17?.

Measurements. – Length 0.51–0.56 mm, height 0.31–0.33 mm.

Remarks. – *Platycythereis* seems to be the most readily available genus for the reception of these thin-shelled juveniles.

The valves are moderately inflated, the shape triangular in lateral aspect. The anterior margin is evenly rounded; the dorsal and ventral margins are straight and converge strongly towards the posterior. The surface is coarsely reticulate, although it is difficult to see the reticula owing to the unremovable matrix. Eye-spots seem to be present.

Genus *Spinoleberis* Deroo, 1966

Spinoleberis kasserinensis Bismuth & Saint-Marc
Pl. 16:4–9

Synonymy. – ☐1980 *Spinoleberis kasserinensis* Bismuth & Saint-Marc – Ben Youssef, pp. 86–87, Pl. 5:2–4, 9–10. ☐1981b *Spinoleberis kasserinensis kasserinensis* n. sp. et n. subsp. Bismuth & Saint-Marc – Bismuth *et al.*, pp. 66–67, Pl. 1:5– 8. ☐1981b *Spinoleberis kasserinensis truncata* n. sp. et n. subsp. Bismuth & Saint-Marc – Bismuth *et al.* p. 67, Pl. 1:9. ☐1981a *Spinoleberis kasserinensis kasserinensis* Bismuth & Saint-Marc; Bismuth *et al.*, p. 235, Pl. 11:13–15. ☐1981a *Spinoleberis kasserinensis truncata* Bismuth & Saint-Marc – Bismuth *et al.*, p. 235. Pl. 11:16. ☐1983 *Spinoleberis kasserinensis kasserinensis* Bismuth & Saint-Marc - Gargouri-Razgallah, pp. 156–157, Pl. 31:3–5. ☐1985 *Spinoleberis kasserinensis* Bismuth & Saint-Marc – Vivière, p. 191, Pl. 12:1–3.

Material. – About 20 carapaces and a few valves.

Occurrence. – Bordj Ghdir section, level A17. Also Djebel Semmama (Lower Cenomanian).

Measurements. – Length 0.59–0.62 mm, height 0.31–0.33 mm.

Description. – The carapace is of medium size, trapezoidal to subtriangular in lateral view. Greatest height is in the plane of the eye-tubercle which is prominent, hemispherical and prolonged ventrally by a vertical ridgelet. The hinge-ear is also prominent. The left valve is larger than the right, overreaching the latter along the major part of the anterior, posterior and ventral margins. However, there tends to be a reversed overhang along the dorsal section of the anterior margin and along the dorsal margin. The anterior margin is evenly rounded in the right valve, more obliquely rounded in the left. There is a marked anteromarginal rim which is sharply delimited from the smooth anterolateral area. On some individuals (mainly those from Algeria), the anteromarginal rim is lined with papillae (see ventral view), whereas others (mainly from Tunisia) are also adorned with stout spines, at least ventrally. There is a slight indentation where the anterior margin joins the feebly concave ventral margin. The dorsal margin is straight. The posterior is characterized by a long, straight dorsal section which reaches far below mid-height; it is rimmed and adorned with variably pronounced spinules, those of the Algerian forms being less prominent.

The posterolateral area is smooth. There is a linking ridgelet between the anteromarginal rim and the ventral ridge. The ventral ridge consists of a curved carina adorned with 6–7 stout tubercles of which the posteroterminal one is particularly prominent. In the mid-dorsal region, there is a flange-like connection between a spiny tubercle and a thick, folded irregular process. The subcentral tubercle is also irregular but distinct and high. The surface between the dorsal and the ventral ridge, including the circumference of the subcentral tubercle, is variably reticulate. The fossae are smoothly rounded. The Algerian representatives are polymorphic with respect to the expression of the reticulation, (e.g. compare Pl. 16:4–5). There is no median ridge, only two isolated spines camouflaged by the surrounding reticulum. A single spine is to be found in the postero- and anteroventral proximity of the subcentral tubercle, also on the anterior side of the dorsal ridge.

The carapace is subhastate and appears particularly spiny in dorsal view. In ventral view, the greatest width lies across the posteroterminal tubercles of the ventral ridge. The ventral surface is reticulate and covered with rounded fossae. Although somewhat recrystallized, the finer details of the hinge appear to agree with *Spinoleberis* (Deroo 1966, p. 165).

Remarks. – As intimated above, the few Algerian representatives differ from those of Djebel Semmama (Tunisia) by their less pronounced marginal spines, polymorphic lateral reticula, and also by a slightly smaller average size.

Bismuth *et al.* (1981b) erected a subspecies *S. kasserinensis truncata*, which was differentiated from *S. kasserinensis kasserinensis* by its more truncated posterior and disconnected mid-dorsal tubercles. As far as can be judged from their figures, the stated differences, at least with respect to the posterior, are slight indeed. The ornamental variability of the Algerian representatives suggests that subspecific separation in this species is unwarranted. Moreover, the biological requirements for the definition of a subspecies, such as geographical and chronological separation are not fulfilled.

Spinoleberis? yotvataensis? Rosenfeld
Pl. 16:10–12

Synonymy. – □?1968 *Cythereis* EmJS 1333 – Grekoff, p. 11, Pl. 1:10a–b. □1973 *Cythereis* EmJS 1333 Grekoff – Bellion *et al.*, pp. 13–14, Pl. 1:6–12. □?1974 *Spinoleberis yotvataensis* n. sp. Rosenfeld – Rosenfeld & Raab, p. 21, Pl. 3:8–11; Pl. 5:11. □1981a *Spinoleberis yotvataensis* Rosenfeld – Bismuth *et al.*, p. 235, Pl. 11:4–6. □?1985 *Spinoleberis yotvataensis macra* n. ssp. – Honigstein & Rosenfeld, p. 454, Pl. 4:3–6. □1985 *Spinoleberis?* gr. *yotvataensis* Rosenfeld – Vivière, pp. 191–193, Pl. 12:4–9.

Material. – A few complete carapaces.

Occurrence. – Somewhat uncertain provenance in the Ras-El-Oued region. Turonian–Coniacian.

Distribution. – Mainly Turonian of Algeria, Tunisia and Israel. For references see synonymy above.

Measurements. – Length males 0.82–0.83 mm, females 0.67–0.68 mm. Height males 0.35–0.37 mm, females 0.36–0.37 mm.

Description. – The surface of the few specimens referred here, is smooth. Sexual dimorphism is pronounced, with the males being longer and somewhat more rectangular in lateral aspect. The anterior margin is broadly and obliquely rounded; the posterior margin has a convex ventral section and a slightly concave dorsal section. Both the anterior and posterior margins are rimmed; the rim of the anterior margin is particularly strong and predominantly smooth. The rudimentary dorsal ridge terminates posteriorly in a pronounced swelling, as does the somewhat more prominent ventral ridge. The subcentral tubercle is prominent, high and isolated from the rudimentary median ridge. The hinge-ear of the left valve is pronounced, overreaching that of the right valve. The eye-tubercle is large but not sharply delimited from the surrounding surface.

Remarks. – As far as can be judged from their figures, more or less identical forms have been reported by Bellion *et al.* (1973) and Bismuth *et al.* (1981a) from the Turonian–Coniacian of Maghreb. They seem to fall within the range of *S. yotvataensis* Rosenfeld, although the latter tends to be larger on average; moreover, there are marked anteomarginal denticles on the Israeli forms. However, a smaller and marginally smoother subspecies, *S. yotvataensis macra* Honigstein & Rosenfeld, has recently been reported from the Late Turonian – Early Coniacian of Israel. The highly polymorphic group of specimens referred to as *Spinoleberis?* gr. *yotvataensis* by Vivière (1985) were subdivided into different groups with respect to ornament. The specimens figured above display the strongest resemblance to those referred to as 'inflated forms' (cf. Vivière 1985, pp. 191–193).

The external habitus of *S. kasserinensis* is significantly different from the forms discussed here, and the concept of the genus *Spinoleberis* is perhaps not wide enough to encompass the latter category conveniently (cf. Deroo 1966, pp. 164–168).

Spinoleberis? sp.
Pl. 16:13–14

Material. – A few complete carapaces.

Occurrence. – Somewhat uncertain provenance in the Ras-El-Oued region. Turonian?

Measurements. – Length 0.67–0.68 mm, height 0.34–0.36 mm.

Remarks. – There is a strong resemblance to *S.? yotvataensis?* with respect to lateral ornament, although our specimens lack the prominent swellings which terminate the dorsal and ventral ridges. The shape, however, is rather different from *S.? yotvataensis?* in that the carapace is more inflated, the ventral margin more convex, and the posterior more rounded. Except for the posterior, the shape agrees quite well with that of *Veenia* or *Veeniacythereis*.

Spinoleberis sp.
Pl. 16:2–3

Material. – A few complete carapaces.

Occurrence. – Bordj Ghdir section, level A10.

Measurements. – Length 0.89–0.91 mm, height 0.49–0.50 mm.

Description. – The carapace is subtrapezoidal in lateral aspect. The hinge-ear is rather pronounced in the left valve as is the eye-tubercle which is surrounded by two distinct notches. The left valve overreaches the right valve along the dorsal part of the anterior margin and along the dorsal section of the posterior margin. The anterior margin, which is rimmed and adorned with spiny denticles (see ventral view), is slightly oblique in the left valve but evenly rounded in the right valve. There is a slight indentation where the anterior margin meets the weakly convex ventral margin which converges towards the posterior with the straight dorsal margin. The posterior is compressed and marginally rimmed; it is convex along its ventral section, but straight along its dorsal section which reaches below mid-height.

The lateral surface is reticulate to punctate, displaying rounded fossae with comparatively broad muri. Being larger in the median region, the fossae become progressively smaller towards the peripheral regions. The antero-lateral area exhibits a few tiny ridgelets. The subcentral tubercle is fairly pronounced and predominantly smooth. There is no median ridge, just a single spine on its site (see ventral view). In the mid-dorsal region, there is a short flange-like connection between a spiny tubercle and a thick, irregular process. The ventral ridge is weakly arched and consists of half a dozen disconnected nodes and a very powerful, posteroterminal, spine (see ventral view). The hastate ventral surface bears two arched ridglets on the anterior two thirds of the surface of each valve. Internal features were not available for study.

Remarks. – The powerful posterior spine on the ventral ridge and the absence of a median ridge strongly support the assignation to *Spinoleberis*.

Genus *Brachycythere* Alexander, 1933

Brachycythere? sp.
Pl. 16:15

Material. – A single carapace.

Occurrence. – Somewhat uncertain provenance in the Ras-El-Oued region. Turonian–Coniacian in age.

Measurements. – Length 0.77 mm, height 0.46 mm.

Remarks. – The shape of this individual agrees with *Brachycythere*. Although some interpretations of *Brachycythere angulata* Grekoff may appear wide enough to encompass the present form (cf. Apostolescu 1961; Honigstein 1984), the single carapace is insufficient for a more precise determination.

Subfamily Buntoniidae Apostolescu, 1961
Genus *Protobuntonia* Grekoff, 1954

Protobuntonia sp.
Pl. 17:1–3

Material. – Half a dozen complete carapaces.

Occurrence. – Somewhat uncertain provenance in the Ras-El-Oued region. Turonian–Coniacian in age.

Measurements. – Length males 0.84–0.85 mm, females 0.72–0.74 mm. Height males 0.43–0.45 mm, females 0.41–0.44 mm.

Remarks. – The external habitus of these specimens resembles *Protobuntonia*. They are strongly dimorphic with respect to length. The longer forms are here interpreted as males; they seem to display the same outline as the significantly smaller Ostracode T1 of Glintzboeckel & Magné (1959). *Protobuntonia numidica* Grekoff, as originaly reported from the Santonian of Algeria, differs from these forms in that the site of maximum height lies further anteriorly. Also, the posterior tends to be more pointed, and the surface more punctate.

Protobuntonia? sp.
Pl. 16:16–17

Material. – A few carapaces.

Occurrence. – Bordj Ghdir section, level A1.

Measurements. – Length 0.38–0.44 mm, height 0.26–0.28 mm.

Description. – A few carapaces of minute size, possibly juveniles, are doubtfully assigned to *Protobutonia*.

The pitted carapace is almond-shaped in lateral aspect, elliptical in dorsal and ventral view. The dorsal margin is weakly convex and decreases markedly in height from front to rear. The anterior is broad and obliquely rounded. There is no genuine anterior cardinal angle at the site of greatest height; there is a posterior cardinal angle. The ventral margin is convex and passes evenly into the posterior in a keel-like compression. There is no eye-tubercle.

Family Hemicytheridae Puri, 1953
Genus *Limburgina* Deroo, 1966

Limburgina sp.
Pl. 17:4–5

Material. – A single adult carapace, and a juvenile.

Occurrence. – Somewhat uncertain provenance in relation to the Bordj Ghdir section.

Measurements. – Length 0.67 mm, height 0.36 mm (adult specimen).

Remarks. – These forms seem to coincide with the properties of *Limburgina* as interpreted by Babinot (1980, pp. 181–183).

The carapace is robust, quadrate and irregularly reticulate in lateral aspect. There are no spines, the dorsal and ventral ridges being smooth, continuous and terminating posteriorly in swellings. A stout rim runs along the broadly rounded anterior and a thinner one along the truncated posterior (damaged on the adult specimen). The subcentral tubercle is flat, rounded and smooth. The eye-tubercle is prominent, the hinge-ear moderately strongly developed.

Order Myodocopida Sars, 1866

Suborder Cladocopa Sars, 1866

Family Polycopidae Sars, 1866

Genus *Polycope* Sars, 1866

Polycope? sp.
Pl. 1:16

Material. – Half a dozen carapaces.

Occurrence. – Bordj Ghdir section, level A14, A15.

Measurements. – Length 0.25–0.26 mm.

Remarks. – This small individual is one of a few recorded with the saucer-like outline usually associated with *Polycope*.

Palaeobiogeography

Initial attempts to chart geographical and/or stratigraphical distributions of fossil species are always more or less provisional, because once more data become available the general picture expands and in some aspects alters as the revision of certain species proceed. The case presented below is no exception to this rule. Apart from the present account, mainly concerned with Albian–Cenomanian species, the largest compilation of palaeogeographical aspects of Cretaceous ostracods from the Tethyan region, has been carried out by Babinot & Colin (1988). Recently, Babinot (1985), Damotte (1985), Vivière (1985), and Athersuch (1988) have also provided important contributions.

(1) The species obtained from sample-levels A7–A17 and available accounts of Upper Albian – Cenomanian ostracods of the Mesogean region and the Middle East (cf. references above), suggest the following 'familiar' subset of species as main elements of a typically Maghrebian (Algerian–Tunisian) association: *Centrocythere tunetana*, *Amphicytherura distincta?*, *Bythocertina tamarae*, *Metacytheropteron bericus*, *Eocytheropteron glintzboeckeli*, *Parexophthalmocythere* sp., *Algeriana cenomanica*, *Cythereis fahrioni*, *Cythereis namousensis*,

Peloriops ziregensis?, *Spinoleberis kasserinensis*, '*Veeniacythereis*' aff. *streblolophata*, '*Veeniacythereis*' *jezzineensis* (Bischoff), *Dolocytheridea atlasica*, *Cytherella gigantosulcata* Rosenfeld and *Perissocytheridea trituberculata* (Rosenfeld) (cf. Glintzboeckel & Magné 1959; Bassoullet & Damotte 1969; Ben Youssef 1980; Gargouri-Razgallah 1983; Vivière 1985). The latter four species have not been recorded in the Bordj Ghdir section.

Although there is a considerable faunal agreement among the various associations reported from the Upper Albian – Cenomanian of Algeria and Tunisia, most of the newly erected species in this monograph are known only from Algeria. Moreover, little is yet known about Morocco, although Reyment (1982) gave a brief review of some Upper Cenomanian – Turonian forms which may possibly display affinities with Algerian and Tunisian species.

(2) Moving east from Maghreb, several of the species occur in Egypt (Colin & El Dakkak 1975), Israel (Rosenfeld & Raab 1974) and Jordan (Babinot & Basha 1985). Some of them, i.e. *Peloriops ziregensis?*, *Cytherella gigantosulcata*, '*Veeniacythereis*' *jezzineensis* and '*Veeniacythereis*' *streblolophata*, are known from Kuwait (Al-Abdul-Razzaq 1977, 1979). The latter three species are also recorded from Iran (Grosdidier 1973), presently the most easterly site yielding typically Maghrebian forms, although recent studies (Athersuch 1988; Babinot & Colin 1988) have extended the records of '*Veeniacythereis*' *jezzineensis* and *Metacytheropteron berbericus* to Oman, Ethiopia and Somalia. Babinot & Colin (1988) mention the occurrence of *Amphicytherura distincta* in Mozambique.

There is consequently a clear resemblance among the various ostracod assemblages all across the North African realm and the Middle East during the Cenomanian. It seems, however, that these affinities are more restricted and superimposed with endemic patterns during the Cenomanian compared with subsequent stages (cf. Babinot 1985; Vivière 1985). An exception to this, is, however, the fading resemblance between the Omanian – East African (Ethiopia, Somalia) and the Maghrebian faunas during Turonian – Early Senonian times as a result of tectonism and differential subsidence in the former region (Athersuch 1988, p. 1197).

A palaeoceanographical reconstruction of Albian sea-currents (Haq 1984), shows a southeast–northwest direction along the African margins. However, it cannot be ascertained (from this investigation) that any relationship exists between currents and ostracod migration along the south shelf of the Tethys sea, although it is interesting to note that '*Veeniacythereis*' *jezzineensis* emerges in the Upper Albian of Lebanon, Israel and Iran, but not until Mid- or Upper Cenomanian in the Maghreb region.

A comparison between the stratigraphical ranges of the more widespread ostracods of the Cenomanian stage of, say, Israel and Maghreb is nevertheless unfeasible due to lack of stratigraphic resolution below the stage level. The following taxa are nevertheless common for the Cenomanian of Maghreb and Israel; *Paracypris dubertreti?*, *Centrocythere tunetana*, *Amphicytherura distincta?*, *Bythoceratina tamarae*, *Metacytheropteron berbericus*, *Cythereis namousensis*, *Dolocythe-*

ridea atlasica, 'Veeniacythereis' jezzineensis and *Peloriops ziregensis.*

(3) According to Vivière (1985), there seems to be a certain agreement between the ostracod assemblages of Maghreb and those of West Africa, particularly Gabon (cf. Grosdidier 1979), at least from the Upper Cenomanian and onwards. This was inferred by, for example, the West African occurrence of the Cenomanian *Perissocytheridea trituberculata, Brachycythere* cf. *sapucariensis* and, later in the Turonian, by *Haughtonileberis mdaourensis* (Bassoullet & Damotte). The trans-Saharan sea-way (Reyment 1980) probably served as a migrational route, as suggested by the distributional pattern of ostracods in the African Paleocene.

(4) Apart perhaps from cytherellids, examples of species in common with Europe are rare indeed. *Metacytheropteron berbericus,* reported from Portugal (Babinot *et al.* 1978), is the only Cenomanian exception in the present collection. Explanations for these faunal dissimilarities, most of which are probably linked to a deep-sea barrier, unfavourable currents and the poor dispersal properties of exclusively benthic organisms, have been discussed in detail by Babinot (1984).

Rather remarkably, the oldest part of the Bordj Ghdir section (levels A1–A6), here estimated as Middle to Late Albian in age, has yielded a small but intriguing collusion consisting of *Matronella* sp. and *Habrocythere* aff. *fragilis,* i.e. species with a marked boreal affinity (Paris basin, Germany; cf. Damotte 1974; Triebel 1940). There is also perhaps a later relict (level A11) of boreal affinity represented by the occurrence of *Eucytherura (Vesticytherura)*? cf. *multituberculata* (cf. Weaver 1982).

However, prior to the manifested Late Albian – Cenomanian segregation between the ostracod faunas of the opposite sides of the Mesogean sea, there are examples of faunal agreements between the two sides (Babinot & Colin 1988). The most convincing example is the *Hechticythere alexanderi – Hechticythere derooi* group which displays a nearly global distribution, having been recorded in southeastern France, Paris Bassin, Great Britain, USA, Venezuela, Ecuador and North Africa (Babinot & Colin 1988, p. 827).

References

Abe, K. 1983: Population structure of *Keijella bisanensis* (Okubo) (Ostracoda, Crustacea). *Journal of the Faculty of Science, University of Tokyo, Section II. 20 (5),* 443–488.

[Al-Abdul-Razzaq, S.Kh. 1977: Study of some Cretaceous Ostracoda of Kuwait. Unpublished thesis, University of Michigan, 1–436 (2 vols).]

Al-Abdul-Razzaq, S.Kh. 1979: *Peloriops,* a new ostracode genus from the Cretaceous of Kuwait. *Proceedings of the Seventh International Symposium on Ostracoda, Beograd (1979),* 47–54.

Al-Abdul-Razzaq, S.Kh. 1983: Biostratigraphic zonation of the Ahmadi formation (Cretaceous, Kuwait) using ostracode assemblages. *In* Maddocks, R.F., (ed): *Applications of Ostracoda,* 394–399. Houston, TX: University of Houston. Department of Geosciences.

Alexander, C.I. 1933: Shell structure of the ostracode genus *Cytheropteron* and fossil species from the Cretaceous of Texas. *Journal of Paleontology 7,* 181–214.

Apostolescu, V. 1961: Contribution à l'étude paléontologique (Ostracodes) et stratigraphique des bassin crétacés et tertiaires de l'Afrique occidentale. *Revue de l'Institut Francais Pétrole 16,* 779–867.

Arkin, V. & Braun, M. 1965: Type sections of Upper Cretaceous formations in the northern Negev (Southern Israel). *Geological Survey of Israel, Stratigraphic Sections, No. 2a.*

Arkin, V., Braun, M. & Starinsky, A. 1965: Type sections of Cretaceous formations in the Jerusalem–Bet Shemesh Area. *Geological Survey of Israel, Stratigraphic Sections, No.1, Part 1, Lithostratigraphy.*

Athersuch, J. 1988: The Biostratigraphy of Cretaceous ostracods from Oman. *In* Hanai, T., Ikeya, N. and Ishizaki, K. (eds.): *Developments in Palaeontology and Stratigraphy 11,* 1187–1206. Proceedings of the Ninth International Symposium on Ostracoda. Kodansha. Elsevier.

[Babinot, J.F. 1980: Les Ostracodes de Crétacé Supérieur du Provence. Thèse, Université de Provence. *Travaux du Laboratoire de Géologie Historique et de Paléontologie 10,* 634 pp. (3 vols.).]

Babinot, J.F. 1984: Quelques réflexions sur la répartition paléobiogéographique des Ostracodes du Crétacé supérieur des marges européenne et nord-africaine. *Travaux du Laboratoire de Stratigraphie et de Paléoecologie, n. ser. 3,* 124–128.

Babinot, J.F. 1985: Paléobiogéographie des Ostracodes du Crétacé supérieur des marges ouest-européennes et nord-africaines de la Téthys. *Bulletin de la Société Géologique de France 8,* 739–745.

Babinot, J.F. & Basha, S.H. 1985: Ostracodes from the Early Cenomanian of Jordan, a preliminary report. *Geobios 18,* 257–262.

Babinot, J.F., Berthou, P.Y., Colin, J.P. & Lauverjat, J. 1978: Les Ostracodes du Cénomanien du Bassin Occidental Portugais: biostratigraphie et affinités paléobiogéographiques. *Cahiers de Micropaleontologie 3,* 11–23.

Babinot, J.F. & Colin, J.P. 1983: Marine Late Cretaceous ostracode faunas from southwestern Europe, a paleoecological synthesis. *In* Maddocks, R.F., (ed.): *Applications of Ostracoda,* 182–205. Department of Geosciences, University of Houston, Houston, Texas.

Babinot, J.F. & Colin, J.P. 1988: Paleobiogeography of Tethyan Cretaceous marine ostracods. *In* Hanai, T., Ikeya, N. & Ishizaki, K., (eds.): *Developments in Palaeontology and Stratigraphy 11,* 823–839. Proceedings of the Ninth International Symposium on Ostracoda. Kodansha. Elsevier.

Bassoullet, J.P. & Damotte, R. 1969: Quelques Ostracodes nouveaux du Cénomanien–Turonien de l'Atlas Saharien Occidentale (Algérie). *Revue de Micropaléontologie 12,* 130–144.

Bellion, Y., Donze, P. & Guiraud, R. 1973: Répartition stratigraphique des principaux Ostracodes (CYTHERACEA) dans le Crétacé supérieur du Sud-Ouest Constantinois (Confins Hodna–Aures, Algérie du Nord). *Publications du Service Géologique de l'Algérie 44,* 7–44.

Bengtson, P. 1988: Open nomenclature. *Palaeontology 31,* 223–227.

[Ben Youssef, M. 1980: Étude stratigraphique et micropaléontologique du Crétacé des Djebels Koumine et Kharroub. Thèse 3ème Cycle, Université de Nice, 104 pp.]

Bischoff, G. 1963: Die Gattung *Cythereis* in der Unterkreide. Ostracoden-Studien im Libanon 1. *Senckenbergiana Lethaea 44,* 1–77.

Bismuth, H., Boltenhagen, C., Donze, P., Le Fèvre, J. & Saint-Marc, P. 1981a: Le Crétacé moyen et supérieur du Djebel Semmama (Tunisie du Centre-Nord); microstratigraphie et évolution sédimentologique. *Bulletin des Centres Recherches Exploration-Production Elf-Aquitaine 5,* 193–267.

Bismuth, H., Donze, P., Le Fèvre, J. & Saint-Marc, P. 1981b: Nouvelles espéces d'Ostracodes dans le Crétacé moyen et supérieur du Djebel Semmama (Tunisie du Centre-Nord). *Cahiers de Micropaleontologie 3,* 51–69.

Bismuth, H., Boltenhagen, C., Donze, P., Le Fèvre, J. & Saint-Marc, P. 1982: Étude sédimentologique et biostratigraphique du Crétacé moyen et supérieur du Djebel Semmama (Tunisie du Centre-Nord). *Cretaceous Research 3,* 171–185.

Cheylan, G., Magné, J., Sigal, J. & Grekoff, N. 1953: Résultats géologiques et micropaléontologique du sondage d'El Krachem

(Hauts plateaux algerois). Description de quelques espèces nouvelles. *Bulletin de la Société Géologique de France, Ser. 6, 3*, 471–492.

Colin, J.P. & El Dakkak, M.W. 1975: Quelques Ostracodes du Cénomanien du Djebel Nezzazat, Sinai, Egypte. *Revista Española de Micropaleontología (nom. especial)*, 49–60.

Cruys, H. 1955: Contributions à l'étude géologique des monts du Hodna, II. La région de Tocqueville et de Bordj-R'dir. *Publications du Service de la Carte Géologique de l'Algérie Bulletin 4*, 195–326.

Damotte, R. 1971: Contribution à l'étude des Ostracodes marins dans le Crétacé du bassin de Paris. *Mémoires de la Société Géologique de France N.S. 113*, 1–150.

Damotte, R. 1974: Un nouveau genre d'Ostracodes de l'Albien du Bassin de Paris: *Matronella* n.g., importance des empreintes musculaires dans les attributions génériques chez les Trachyleberidinae. *Palaeontographica, Abt. A, 146*, 181–189.

Damotte, R. 1977: Sur les genres crétacés: *Rehacythereis, Veeniacythereis, Cornicythereis, Parvacythereis* and *Chapmanicythereis* (Trachyleberididae, Ostracoda) créés par J. Gründel en 1973. *Revue de Micropaléontologie 19*, 200–210.

Damotte, R. 1985: Les Ostracodes du Crétacé moyen sud-mésogéen et leur répartition paléogéographique. *Bulletin de la Société Géologique de France 8*, 733–737.

Damotte, R. & Grosdidier, E. 1963: Quelques Ostracodes du Crétacé inférieur de la Champagne Humide 2. – Aptien. *Revue de Micropaléontologie 6*, 153–168.

Damotte, R. & Rey, J. 1980: Ostracodes du Crétacé inférieur d'Estremadura (Portugal). *Revue de Micropaléontologie 23*, 16–36.

Damotte, R. & Saint-Marc, P. 1972: Contribution à la connaissance des Ostracodes crétacés du Liban. *Revista Española de Micropaleontología 4*, 273–296.

Deroo, G. 1956: Études critiques au sujet des Ostracodes marins du Crétacé inférieur et moyen de la Champagne Humide et du Bas Boulonnais. *Revue de l'Institut Français du Pétrole 11*, 1499–1535.

Deroo, G. 1966: Cytheracea (Ostracodes) du Maastrichtien de Maastricht (Pays-Bas) et de régions voisines: résultats stratigraphiques et paléontologiques de leur étude. *Mededelingen van de Geologische Stichting C2*, 197 pp.

Donze, P. & Porthault, B. 1972: Les Ostracodes de la sous-famille des Trachyleberidinae dans quelques coupes de reference du Cénomanien du sud-est de la France. *Revista Española de Micropaleontología 4*, 355–376.

Donze, P. & Thomel, G. 1972: Le Cénomanien de La Foux (Alpes de Haute-Provence) Biostratigraphie et faunes nouvelles d'Ostracodes. *Eclogae Geologica Helvetica 65*, 369–389.

Ducasse, O. 1981: Étude populationniste du genre *Cytherella* (Ostracodes) dans la facies du paléogène aquitain. Intéret dans la reconstitution des paléoenvironments. *Bulletin de l'Institut Géologie du Bassin d'Aquitaine 30*, 161–186.

Esker, G.C. 1968: Danian Ostracods from Tunisia. *Micropaleontology 14*, 319–333.

[Gargouri-Razgallah, S. 1983: Le Cénomanien de Tunisie centrale. Étude paléoécologique, stratigraphique, micropaléontologique et paléogéographique. (Thèse) *Documents et Travaux, IGAL, Paris. 6*.]

Gerry, E. & Rosenfeld, A. 1973: *Amphicytherura distincta* and *Neocyprideis vandenboldi* (Ostracoda), new species of the Cenomanian–Turonian of Israel. *Revista Española de Micropaleontología 5*, 99–105.

Glintzboeckel, C. & Magné, J. 1959: Répartition des microfaunes à plancton et à Ostracodes dans le Crétacé supérieur de la Tunisie et de l'Est algérien. *Revue de Micropaléontologie 2*, 57–67.

Grekoff, N. 1969: Sur la valeur stratigraphique et les relations paléogéographique de quelques Ostracodes du Crétacé, du Paléocène et de l'Eocène inférieur d'Algérie orientale. *Proceedings of the third African Micropaleontological Colloqium, Cairo, 1969*, 227–248.

Grosdidier, E. 1973: Associations d'Ostracodes du Crétacé d'Iran. *Revue de l'Institut Français du Pétrole 28*, 131–168.

Grosdidier, E. 1979: Principaux ostracodes marins de l'intervalle Aptien–Turonien du Gabon (Afrique occidentale). *Bulletin des Centres de Recherches Exploration-Production Elf-Aquitaine 3*, 1–35.

Gründel, J. 1973: Zur Entwicklung der Trachyleberididae (Ostracoda) in der Unterkreide und in der tieferen Oberkreide. Teil I: Taxonomie. *Zeitschrift Geologischer Wissenschaften 1*, 1463–1474.

Gründel, J 1974: Zur Entwicklung der Trachyleberididae (Ostracoda) in der Unterkreide und in der tieferen Oberkreide. Teil II: Phylogenie. *Zeitschrift Geologischer Wissenschaften 2*, 61–71.

Haq, B.U. 1984: Paleoceanography: a synoptic overview of 200 million years of ocean history. *In* Haq, B.V. & Milliman, J.D. (eds.): *Marine geology and oceanography of Arabian Sea and coastal Pakistan*, 201–231. Van Nostrand Reinhold Co., N.Y.

Hart, M.B. 1980: The recognition of Mid-Cretaceous sea-level changes by means of foraminifera. *Cretaceous Research 1*, 289–297.

Hartmann, G. 1966: Ostracoda. Dr. H.G. Bronn, *Klassen und Ordnungen des Tierreichs*. Fünfter Band: Arthropoda, I. Abteilung: Crustacea. 2 Buch, IV. Teil. VEB Gustav Fischer Verlag, Jena.

Hartmann, G. & Puri, H.S. 1974: Summary of neontological and paleontological classification of Ostracoda. *Mitteilungen aus dem Hamburgischen zoologischen Museum und Institut 70*, 7–73.

Hill, B.L. 1954: Reclassification of winged *Cythereis* and winged *Brachycythere*. *Journal of Paleontology 28*, 804–826.

Honigstein, A. 1984: Senonian ostracodes from Israel. *Israel Geological Survey, Bulletin 78*, 1–48.

Honigstein, A. & Rosenfeld, A. 1985: Late Turonian – Early Coniacian ostracodes from the Zihor Formation, southern Israel. *Revista Española de Micropaleontología 17*, 447–466.

Kaye, P. 1964: Revision of British marine Cretaceous Ostracoda with notes on additional forms. *Bulletin of the British Museum (Nat. Hist.) Geology 10*, 35–79.

[Liebau, A. 1971: Homologe Skulpturmuster bei Trachyleberididen und verwandten Ostrakoden. Dissertation, Universität Berlin.]

Majoran, S. 1987a: On *Maghrebeis tuberculata* gen. et sp. nov. *A Stereo-Atlas of Ostracod Shells 14:1*, 29–32.

Majoran, S. 1987b: Notes on mid-Cretaceous biostratigraphy of Algeria. *Journal of African Earth Sciences 6*, 781–786.

Majoran, S. 1988: Comments on a miscellaneous ostracod group from the mid-Cretaceous of the south shelf of the Tethys sea. *Journal of African Earth Sciences 7*, 691–702.

Mayr, E. 1963: *Animal Species and Evolution*. Belknap Press, Harvard University, Cambridge, MA.

Morkhoven, F.P.C.M. Van 1962: *Post-Palaeozoic Ostracoda: Their Morphology, Taxonomy and Economic Use 1*, 1–244. Elsevier, Amsterdam.

Morkhoven, F.P.C.M. Van 1963: *Post-Palaeozoic Ostracoda: Their Morphology, Taxonomy and Economic Use 2*, 1–478. Elsevier, Amsterdam.

Neale, J.W. 1978: The Cretaceous. *In* Bate, R.H. & Robinson, E. (eds.): *A stratigraphical index of British Ostracoda. Geological Journal Special Issue 8*, 325–384.

Neale, J.W. 1982: Aspects of the Subfamily Schulerideinae. *In* Bate, R.H., Robinson, E. & Sheppard, L.M. (eds.): *Fossil and Recent Ostracods*, 178–192. British Micropalaeontological Society Series. Ellis Horwood, Chichester.

Reyment, R.A. 1980: Biogeography of the Saharan Cretaceous and Paleocene epicontinental transgressions. *Cretaceous Research 1*, 299–327.

Reyment, R.A. 1982: Note on Upper Cretaceous Ostracods from South-western Morocco. *Cretaceous Research 3*, 405–414.

Reyment, R.A. 1984: Upper Cretaceous Ostracoda of North Central Spain. *Bulletin of the Geological Institutions of the University of Uppsala N. S. 10*, 67–110.

Reyment, R.A. 1985: Phenotypic evolution in a lineage of the Eocene ostracod *Echinocythereis*. *Paleobiology 11*, 174–194.

Reyment, R.A. (in press.): On polymorphism in a variable environment, with application to marine ostracods. *Proceedings of the Tenth International Symposium on Ostracoda. Aberystwyth. 1988*.

[Rodriguez Lazaro, J.M. 1985: Los Ostracodos del Coniaciense y Santoniense de la Cuenca Vasco-Cantábrica Occidental. Tesis. Facultad de Ciencias Universidad del Pais Vasco.]

Rosenfeld, A. & Raab, M. 1974: Cenomanian–Turonian ostracodes from the Judea Group in Israel. *Israel Geological Survey, Bulletin 62*, 1–64.

Rosenfeld, A. & Raab, M. 1984: Lower Cretaceous ostracodes from Israel and Sinai. *Israel Journal of Earth-Sciences 33*, 85–134.

Szczechura, J. 1971: Seasonal changes in a reared fresh-water species, *Cyprinotus (Heterocypris) incongruens* (Ostracoda), and their importance in the interpretation of variability in fossil ostracodes. *Bulletin du Centre Recherches de Pau–SNPA 5 (supplement)*, 191–205.

Sylvester-Bradley, P.C. & Benson, R.H. 1971: Terminology for surface features in ornate ostracodes. *Lethaia 4*, 249–286.

Triebel, E. 1940: Die Ostracoden der deutschen Kreide. *Senckenbergiana 22*, 160–227.

Van Veen, J.E. 1932: Die Cytherellidae der Maastrichter Tuffkreide und des Kunrader Korallenkalkes von Süd-Limburg. *Overgedrukt uit de Verhandelingen van het Geologisch-Mijnbouwkundig Genootschap voor Nederland en Kolonien 9*, 317–364.

[Vivière, J.L. 1985: Les ostracodes du Crétacé supérieur (Vraconien à Campanien basal) de la région de Tébessa (Algérie du Nord-Est): Stratigraphie, Paléoécologie, Systématique. Thèse 3 cycle, Académie de Paris Université Pierre et Marie Curie. *Mémoires des Sciences de la Terre, Paris VI*, 1–261.]

Weaver, P.P.E. 1982: Ostracoda from the British Lower Chalk and plenus marls. *Palaeontological Society, Monograph 135*, 1–127.

Whatley, R. 1983: The application of Ostracoda to palaeoenvironmental analysis. *In* Maddocks, R.F. (ed.): *Applications of Ostracoda*, 51–77. Department of Geosciences, University of Houston, Houston, Texas.

Plates

Plate 1

☐1. *Cytherella* cf. *ovata* Roemer. Right side of a female carapace. Bordj Ghdir section, level A14. ×68. PMAL30. [Page 8.]

☐2. Same species and provenance. Left side of a female carapace. ×68. PMAL31.

☐3. Same species and provenance. Female carapace in dorsal view. ×68. PMAL32.

☐4. Same species; Bordj Ghdir section, level A12. Right side of a male carapace. ×68. PMAL33.

☐5. Same species and provenance. Left side of a male carapace. ×68. PMAL34.

☐6. Same species and provenance. Male carapace in dorsal view. ×68. PMAL35.

☐7. Same species? (A-1?); Bordj Ghdir section, level A2. Right side of a male carapace. ×68. PMAL36.

☐8. Same species? (A-1?) and provenance. Left side of a carapace. ×68. PMAL37.

☐9. *Cytherella* aff. *contracta* Van Veen. Left side of a carapace. Bordj Ghdir section, level A2. ×68. PMAL38. [Page 8.]

☐10. Same species; Bordj Ghdir section, level A14. Left side of a carapace. ×68. PMAL39.

☐11. *Cytherella sulcata?* Van Veen. Left side of a carapace. Bordj Ghdir section, level A12. ×68. PMAL40. [Page 8.]

☐12. Same species and provenance. Right side of a carapace. ×68. PMAL41.

☐13. Same species and provenance. Carapace in dorsal view. ×68. PMAL42.

☐14. Same species. Right side of a carapace. Bordj Ghdir section, level A2. ×68. PMAL43.

☐15. Same species and provenance. Left side of a carapace. ×68. PMAL44.

☐16. *Polycope?* sp. Carapace. Bordj Ghdir section, level A14. ×136. PMAL45. [Page 29.]

Plate 2

☐1. *Bairdia* spp. Right valve. Bordj Ghdir section, level A2. ×51. PMAL46. [Page 9.]

☐2. Same species? Left valve. Bordj Ghdir section, level A4. ×51. PMAL47.

☐3. Same species? Right valve. Bordj Ghdir section, level A4. ×51. PMAL48.

☐4. Same species? Right valve. Bordj Ghdir section, level A5. ×51. PMAL49.

☐5. *Bairdia* sp. Right side of a carapace. Bordj Ghdir section, level A14. ×51. PMAL50. [Page 9.]

☐6. Same species and provenance. Carapace in dorsal view. ×51. PMAL51.

☐7. Same species and provenance. Left side of a carapace. ×51. PMAL52.

☐8. *Bythocypris? symmetrica* sp. nov. Carapace in dorsal view. Bordj Ghdir section, level A10. ×102. PMAL53. [Page 9.]

☐9. Same species and provenance. Left side of a carapace. ×68. PMAL54.

☐10. *Paracypris dubertreti?* Damotte & Saint-Marc. Right side of a carapace. Bordj Ghdir section, level A13. ×68. PMAL55. [Page 10.]

☐11. Same species and provenance. Carapace in dorsal view. ×68. PMAL56.

☐12. Same species and provenance. Left side of a carapace. ×68. PMAL57.

☐13. *Paracypris?* sp. Right side of a carapace. Bordj Ghdir section, level A2. ×68. PMAL58. [Page 10.]

☐14. Same species and provenance. Carapace in dorsal view. ×68. PMAL59.

☐15. Same species and provenance. Left side of a carapace. ×68. PMAL60.

Plate 3

☐1. *Pontocyprella maghrebensis* sp. nov. Left side of a carapace. Bordj Ghdir section, level A14. ×68. PMAL61. [Page 10.]

☐2. Same species and provenance. Carapace in dorsal view. ×68. PMAL62.

☐3. Same species and provenance. Right side of a carapace. ×68. PMAL63.

☐4. *Macrocypris?* sp. Left side of a carapace. Bordj Ghdir section, level A10. ×68. PMAL64. [Page 10.]

☐5. *Dolocytheridea?* sp. Left side of a male carapace. Road-section, level A17. ×68. PMAL65. [Page 11.]

☐6. Same species and provenance. Ventral view of a female carapace. ×68. PMAL66.

☐7. Same species and provenance. Right side of a male carapace. ×68. PMAL67.

☐8. Same species and provenance. Left side of a female carapace. ×68. PMAL68.

☐9. Same species and provenance. Right side of a female carapace. ×68. PMAL69.

☐10. *'Dolocytheridea' polymorphica* sp. nov. Right valve of a male. Bordj Ghdir section, level A16. ×136. PMAL70. [Page 11.]

☐11. Same species and provenance. Right valve of a male. ×136. PMAL71.

☐12. Same species and provenance. Left valve of a male. ×102. PMAL72.

☐13. Same species and provenance. Left valve of a female. ×102. PMAL73.

☐14. *Habrocythere* aff. *fragilis* Triebel. Right side of a carapace. Bordj Ghdir section, level A2. ×136. PMAL74. [Page 11.]

☐15. Same species and provenance. Left side of a carapace. ×136. PMAL75.

Plate 4

☐1. *Centrocythere tunetana* Bismuth & Donze. Left side of a male? carapace. Bordj Ghdir section, level A12. ×94. PMAL76. [Page 11.]

☐2. Same species and provenance. Right side of a male? carapace. ×94. PMAL77.

☐3. Same species from Djebel Semmama. Showing 'outgrowths' on the reticulum walls. ×238. PMTN11.

☐4. Same species from the Bordj Ghdir section, level A12. Reticulum walls. ×340. PMAL78.

☐5. Same species from Djebel Semmama. Hinge of right valve. ×170. PMTN12.

☐6. *Amphicytherura distincta*? Gerry & Rosenfeld. Left side of a carapace. Bordj Ghdir section, level A4. ×136. PMAL79. [Page 12.]

☐7. Same species and provenance. Right side of a carapace. ×136. PMAL80.

☐8. Same species from the Bordj Ghdir section, level A16. Left side of a carapace. ×136. PMAL81.

☐9. Same species and provenance. Right side of a carapace. ×136. PMAL82.

☐10. *Eucythere*? sp. Left valve. Bordj Ghdir section, level A14. ×102. PMAL83. [Page 12.]

☐11. Same species and provenance. Left valve. ×102. PMAL84.

☐12. Same species and provenance. Right valve. ×102. PMAL85.

☐13. Same species and provenance. Carapace in ventral view. ×102. PMAL86.

☐14. Same species; Bordj Ghdir section, level A16. Left side of a carapace. ×136. PMAL87.

☐15. Same species and provenance. Right valve. ×102. PMAL88.

Plate 5

☐1. *'Eucythere' mackenziei* sp. nov. Left valve. Bordj Ghdir section, level A16. ×94. PMAL89. [Page 13.]

☐2. Same species and provenance. Left side of a carapace. ×94. PMAL90.

☐3. Same species and provenance. Right valve. ×94. PMAL91.

☐4. Same species and provenance. Left valve. ×94. PMAL92.

☐5. Same species and provenance. Internal view of left valve. ×94. PMAL93.

☐6. *Bythoceratina tamarae* Rosenfeld. Right valve. Bordj Ghdir section, level A15. ×102. PMAL94. [Page 13.]

☐7. Same species and provenance. Left valve. ×102. PMAL95.

☐8. Same species from Djebel Semmama. Right side of a carapace. ×102. PMTN13.

☐9. Same species and provenance. Carapace in dorsal view. ×102. PMTN14.

☐10. *Monoceratina? hodnaensis* sp. nov. Right side of a carapace. Bordj Ghdir section, level A13. ×102. PMAL96. [Page 13.]

☐11. Same species and provenance. Left valve. ×102. PMAL97.

☐12. Same species and provenance. Carapace in dorsal view. ×102. PMAL98.

Plate 6

☐1. *Metacytheropteron berbericus* Bassoullet & Damotte. Right side of a carapace. Bordj Ghdir section, level A17. ×136. PMAL99. [Page 14.]

☐2. Same species from the Qastel (Israel, Jerusalem). Right side of a carapace. ×136. PMIsr11.

☐3. *Metacytheropteron* sp. A. Right valve. Bordj Ghdir section, level A1. ×102. PMAL100. [Page 14.]

☐4. Same species and provenance. Right valve. ×102. PMAL101.

☐5. *Eocytheropteron?* sp. Right side of a carapace. Bordj Ghdir section, level A14. ×136. PMAL102. [Page 15.]

☐6. *Metacytheropteron* sp. B. Left valve. Bordj Ghdir section, level A2. ×68. PMAL103. [Page 14.]

☐7. Same species and provenance. Right valve. ×68. PMAL104.

☐8. *Cytheropteron?* sp. Carapace in dorsal view. Bordj Ghdir section, level A16. ×136. PMAL106. [Page 14.]

☐9. Same species and provenance. Right side of a carapace. ×136. PMAL107.

☐10. Same species and provenance. Left side of a carapace. ×136. PMAL108.

☐11. *Eocytheropteron glintzboeckeli* Donze & Lefèvre. Hinge of left valve. Djebel Semmama. ×136. PMTN15. [Page 15.]

☐12. Same species from the Bordj Ghdir section, level A14. Right valve. ×102. PMAL109.

☐13. Same species and provenance. Left valve. ×102. PMAL110

Plate 7

☐1. *'Cytherura'* sp. Left side of a carapace. Bordj Ghdir section, level A2. ×136. PMAL111. [Page 15.]

☐2. *Eucytherura* (*Vesticytherura*)? sp. Left side of a carapace. Bordj Ghdir section, level A2. ×136. PMAL112. [Page 16.]

☐3. *'Procytherura'* sp. Right side of a carapace. Bordj Ghdir section, level A2. ×136. PMAL113. [Page 16.]

☐4. Same species and provenance. Left side of a carapace. ×136. PMAL114.

☐5. *Procytherura? cuneata* sp. nov. Left side of a carapace. Road-section, level A2. ×136. PMAL115. [Page 15.]

☐6. Same species and provenance. Right side of a carapace. ×102. PMAL116.

☐7. *Eucytherura* (*Vesticytherura*)? cf. *multituberculata* Gründel. Left side of a carapace. Bordj Ghdir section, level A11. ×153. PMAL117. [Page 16.]

☐8. Same species and provenance. Internal view of left valve ×153. PMAL118.

☐9. *Parexophthalmocythere* sp. Left side of a carapace. Bordj Ghdir section, level A13. ×73. PMAL119. [Page 17.]

☐10. Same species and provenance. Right side of a carapace. ×73. PMAL120.

☐11. Same species; Bordj Ghdir section, level A10. Left side of a carapace. ×73. PMAL121.

☐12. Same species?; Bordj Ghdir section, level A7. Right side of a carapace. ×73. PMAL122.

☐13. *Pterygocythere* sp. Ventral view of a carapace. Bordj Ghdir section, level A10. ×51. PMAL123. [Page 17.]

☐14. Same species and provenance. Carapace in dorsal view. ×51. PMAL124.

☐15. Same species and provenance. Right side of a carapace. ×51. PMAL125.

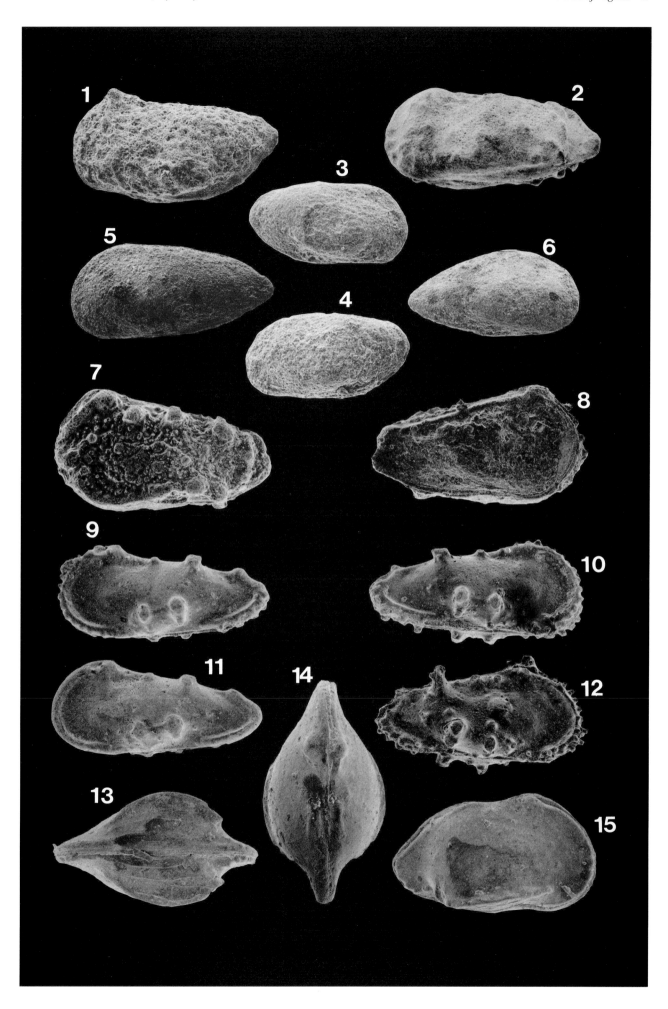

Plate 8

☐1. *Algeriana reymenti* gen. et sp. nov. Left side of a male carapace. Bordj Ghdir section, level A12. ×51. PMAL126. [Page 18.]

☐2. Same species and provenance. Right side of a male carapace. ×51. PMAL127a.

☐3. Same species; Bordj Ghdir section, level A14. Internal view of a male, right valve. ×51. PMAL127b.

☐4. Same species and provenance. Internal view of a male, left valve. ×51. PMAL128.

☐5. Same species; Bordj Ghdir section, level A12. Male carapace in ventral view. (stereographs). ×51. PMAL129.

☐6. Same species and provenance. Dorsal view of a male carapace. ×51. PMAL130.

☐7. Same species and provenance. Female carapace in dorsal view. ×51. PMAL131.

☐8. Same species and provenance. Female carapace in ventral view. ×51. PMAL132.

☐9. Same species and provenance. Left side of a female carapace. ×51. PMAL133.

☐10. Same species and provenance. Right side of a female carapace. ×51. PMAL134.

☐11. Same species; Bordj Ghdir section, level A14. Internal view of a female, right valve. ×51. PMAL135.

☐12. Same species; Bordj Ghdir section, level A12. Right side of a carapace (A-1). ×51. PMAL136.

☐13. Same species; Bordj Ghdir section, level A13. Showing muscle-scars. ×510. PMAL137.

☐14. Same species. Anterior hinge-tooth of the right valve. ×340. PMAL138.

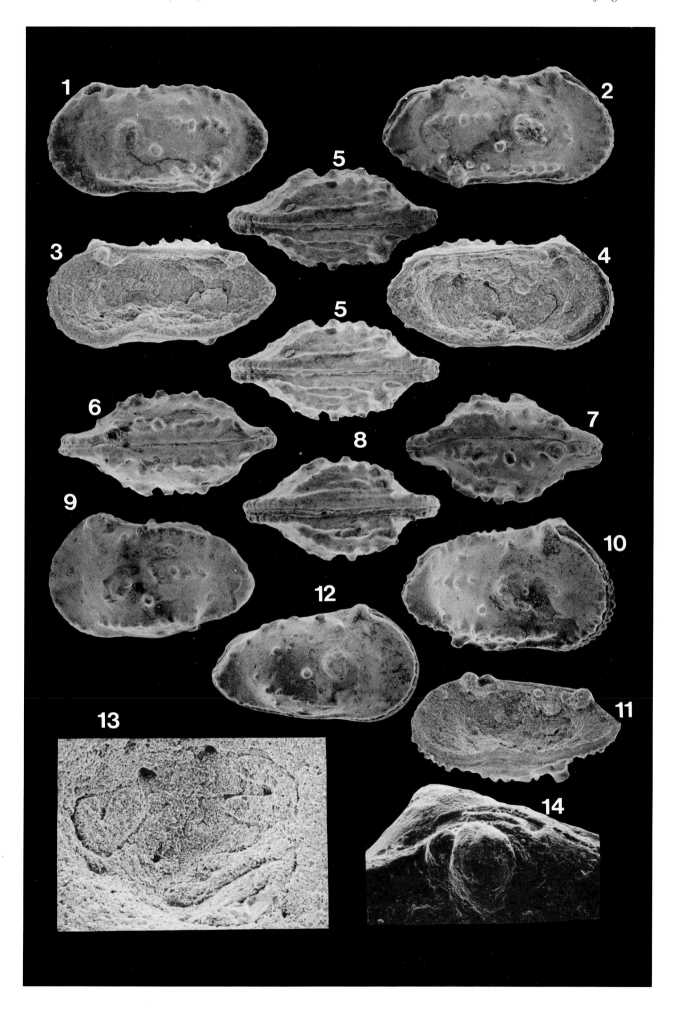

Plate 9

☐1. *Algeriana cenomanica* gen. et sp. nov. Left side of a male carapace. Djebel Semmama. ×68. PMTN15. [Page 18.]

☐2. Same species and provenance. Right side of a male carapace. ×68. PMTN16.

☐3. Same species; Bordj Ghdir section, level A17. Right side of a female carapace. ×68. PMAL139.

☐4. Same species; Bordj Ghdir section, level A17. Right side of a female carapace. ×68. PMAL140.

☐5. Same species; Bordj Ghdir section, level A17. Right side of a male carapace. ×68. PMAL141.

☐6. Same species from Djebel Semmama. Ventral view of a male carapace. ×68. PMTN17.

☐7. Same species and provenance. Left side of a female carapace. ×102. PMTN18.

☐8. Same species and provenance. Dorsal view of a female carapace. ×128. PMTN19.

☐9. Same species and provenance. Muscle-scars. ×510. PMTN20.

☐10. Same species and provenance. Internal view of a damaged, left valve. ×255. PMTN21.

☐11. Same species and provenance. Hinge of a damaged right valve. ×102. PMTN22.

☐12. Same species and provenance. Hinge of a damaged right valve. ×136. PMTN23.

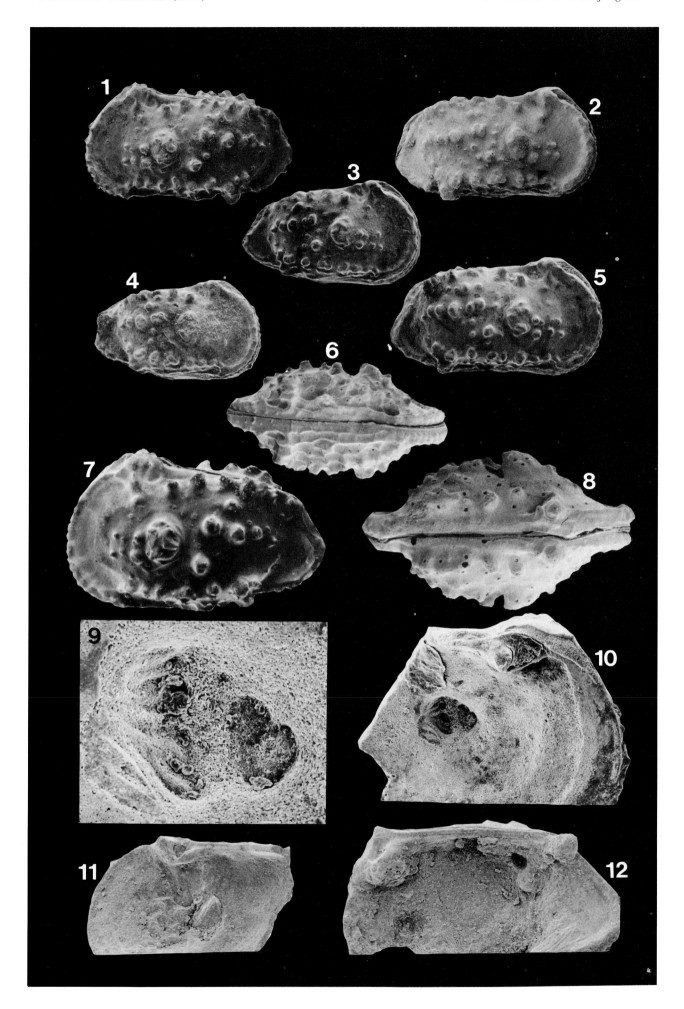

Plate 10

☐1. *'Cythereis'* sp. Male carapace in ventral view. Bordj Ghdir section, level A2. ×64. PMAL142. [Page 19.]

☐2. Same species and provenance. Left side of a male carapace. ×64. PMAL143.

☐3. Same species and provenance. Right side of a female carapace. ×64. PMAL144.

☐4. Same species and provenance. Female carapace in dorsal view. ×64. PMAL145.

☐5. *Cythereis* aff. *fahrioni* Bischoff. Left side of a carapace. Bordj Ghdir section, level A2. ×66. PMAL146. [Page 19.]

☐6. Same species and provenance. Right side of a carapace. ×66. PMAL147.

☐7. Same species and provenance. Carapace in dorsal view. ×68. PMAL148.

☐8. Same species and provenance. Carapace in ventral view. ×68. PMAL149.

☐9. Same species and provenance. Carapace in ventral view. ×68. PMAL150.

☐10. Same species and provenance. Carapace in dorsal view. ×68. PMAL151.

☐11. Same species and provenance. Left side of a carapace. ×68. PMAL152.

☐12. Same species and provenance. Left side of a carapace. ×68. PMAL153.

☐13. *Cythereis namousensis* Bassoullet & Damotte. Left valve (A-1). Bordj Ghdir section, level A16. ×102. PMAL154. [Page 21.]

☐14. Same species and provenance. Left valve (A-1). ×85. PMAL155.

☐15. Same species from the Qastel (Israel). Left side of a carapace. ×68. PMIsr12.

☐16. Same species from the Qastel. Right side of a carapace. ×68. PMIsr13.

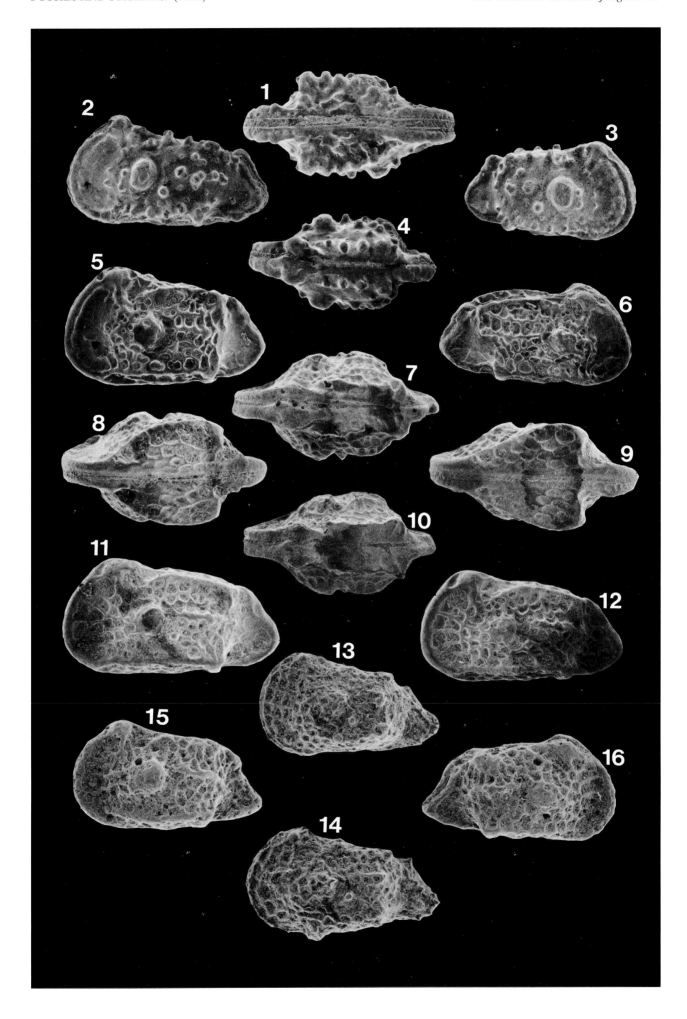

Plate 11

☐1. *Cythereis fahrioni bigrandis* subsp. nov. Left side of a carapace, small specimen. Djebel Semmama. ×102. PMTN24. [Page 20.]

☐2. Same species and provenance. Right side of a carapace, small specimen. ×102. PMTN25.

☐3. Same species and provenance. Carapace in ventral view, small specimen. ×102. PMTN26.

☐4. Same species and provenance. Left side of a carapace, small specimen. ×102. PMTN27.

☐5. Same species and provenance. Left side of a carapace, small specimen. ×102. PMTN28.

☐6. Same species and provenance. Carapace in dorsal view, small specimen. ×102. PMTN29.

☐7. Same species and provenance. Male carapace in dorsal view, large specimen. ×68. PMTN30.

☐8. Same species and provenance. Female carapace in dorsal view, large specimen. ×68. PMTN31.

☐9. Same species and provenance. Left side of a female carapace, large specimen. ×102. PMTN32.

☐10. Same species and provenance. Left side of a male carapace, large specimen. ×68. PMTN33.

☐11. Same species and provenance. Right side of a male carapace, large specimen. ×68. PMTN34.

☐12. Same species and provenance. Muscle-scars. ×680. PMTN35.

☐13. Same species and provenance. Anterior hinge-tooth of a right valve. ×510. PMTN36.

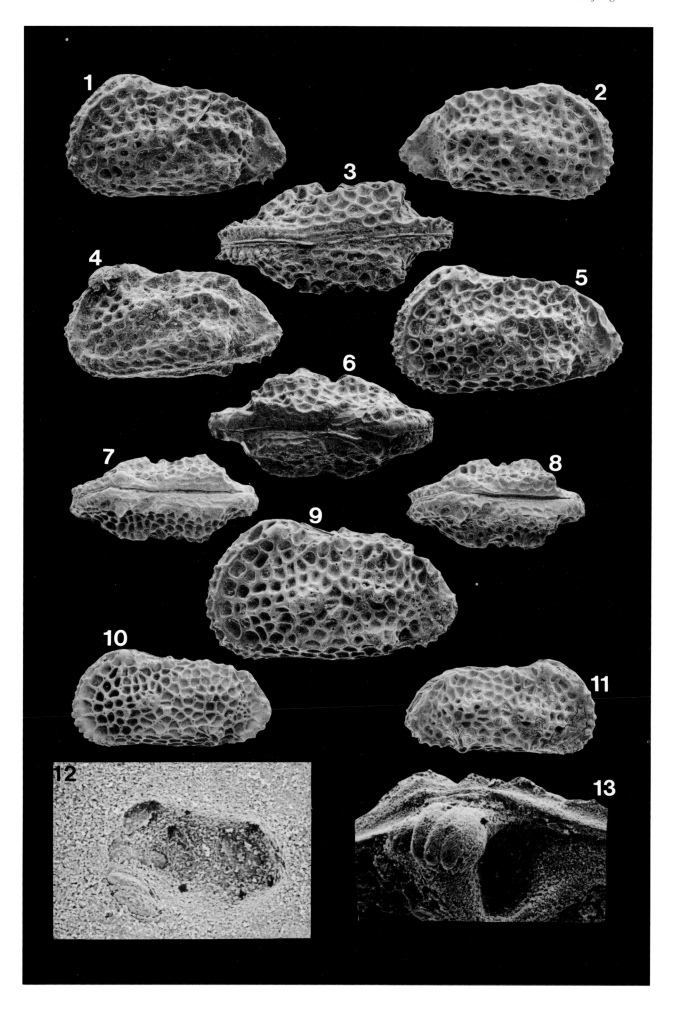

Plate 12

☐1. *Cythereis fahrioni* Bischoff. Left side of a male carapace. Bordj Ghdir section, level A15. ×68. PMAL156. [Page 20.]

☐2. Same species and provenance. Right side of a female carapace. ×102. PMAL157.

☐3. Same species, somewhat uncertain provenance. Left side of a female carapace. ×102. PMAL158.

☐4. Same species and provenance. Left side of a male carapace. ×68. PMAL159.

☐5. *Cythereis? punctatafoveolata* sp. nov. Left side of a female carapace. Bordj Ghdir section, level A15. ×74. PMAL160. [Page 21.]

☐6. Same species and provenance. Right side of a female carapace. ×73. PMAL161.

☐7. Same species and provenance. Female carapace in ventral view. ×73. PMAL162.

☐8. Same species and provenance. Left side of a male carapace. ×73. PMAL163.

☐9. Same species and provenance. Right side of a male carapace. ×73. PMAL164.

☐10. Same species and provenance. Male carapace in ventral view. ×73. PMAL165.

☐11. Same species and provenance. Right side of a male carapace. ×73. PMAL166.

☐12. Same species and provenance. Internal view of right valve. ×170. PMAL167.

☐13. Same species and provenance. Posterior hinge-tooth of right valve. PMAL168.

Plate 13

☐1. *Cythereis afroreticulata* sp. nov. Left valve. Bordj Ghdir section, level A14. ×51. PMAL169. [Page 22.]

☐2. Same species and provenance. Right valve. ×47. PMAL170.

☐3. Same species; Bordj Ghdir section, level A16. Carapace in dorsal view. ×51. PMAL171.

☐4. Same species; Bordj Ghdir section, level A13. Left valve of an instar (A-1). ×61. PMAL172.

☐5. Same species and provenance. Left valve of an instar (A-1). ×68. PMAL173.

☐6. Same species; Bordj Ghdir section, level A16. Carapace in ventral view. ×47. PMAL174.

☐7. Same species; Bordj Ghdir section, level A14. Hinge of right valve. PMAL175.

☐8. Same individual. ×51.

☐9. *Veeniacythereis?* sp. Right side of a carapace. Bordj Ghdir section level A7. ×62. PMAL176. [Page 23.]

☐10. Same species? of somewhat uncertain provenance. Left side of a carapace. ×68. PMAL177.

☐11. Same species? of somewhat uncertain provenance. Carapace in dorsal view. ×68. PMAL178.

☐12. *'Veeniacythereis' subrectangulata* Majoran. Left side of a female carapace. Bordj Ghdir section, level A10. ×63. PMAL11. [Page 23.]

☐13. Same species and provenance. Right side of a female carapace. ×63. PMAL179.

☐14. Same species and provenance. Male carapace in dorsal view. ×51. PMAL180.

☐15. Same species and provenance. Male carapace in ventral view. ×51. PMAL181.

☐16. Same species and provenance. Right side of a male carapace. ×51. PMAL182.

☐17. *'Veeniacythereis'* sp. nov? Right side of a male carapace. Djebel Semmama. ×68. PMTN37. [Page 23.]

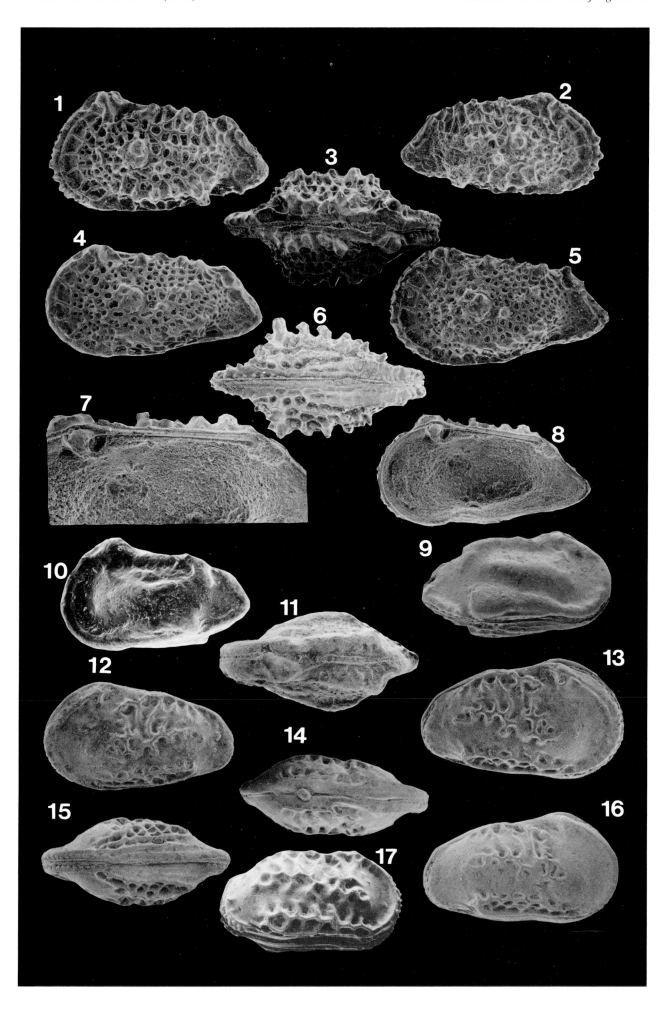

Plate 14

☐1. *Maghrebeis tuberculata* Majoran. Left side of a carapace (stereographs). Bordj Ghdir section, level A16. ×111. PMAL183. [Page 23.]

☐2. Same species and provenance. Carapace in ventral view (stereographs). ×111. PMAL5.

☐3. Same species and provenance. Right side of a carapace (stereographs). ×111. PMAL6.

☐4. *Peloriops ziregensis*? (Bassoullet & Damotte). Female carapace in ventral view. Bordj Ghdir section, level A15. ×77. PMAL184a. [Page 24.]

☐5. Same species and provenance. Male carapace in dorsal view. ×63. PMAL184b.

☐6. Same species from Djebel Semmama. Right side of a female carapace (stereographs). ×73. PMTN38.

☐7. Same species from Djebel Semmama. Right side of a male carapace (stereographs). ×68. PMTN39.

☐8. Same species; Bordj Ghdir section, level A12. Left side of a female carapace. ×70. PMAL185.

☐9. Same species; Bordj Ghdir section, level A15. Left side of a male carapace. ×64. PMAL186.

☐10. Same species; Bordj Ghdir section, level A12. Right side of a male carapace. ×60. PMAL187.

☐11. Same species and provenance. Right side of a female carapace. ×77. PMAL188.

☐12. Same species from Djebel Semmama. Anterior hinge-tooth of right valve. ×102. PMTN40.

☐13. Same species from Djebel Semmama. Posterior hinge-tooth of right valve. ×180. PMTN41.

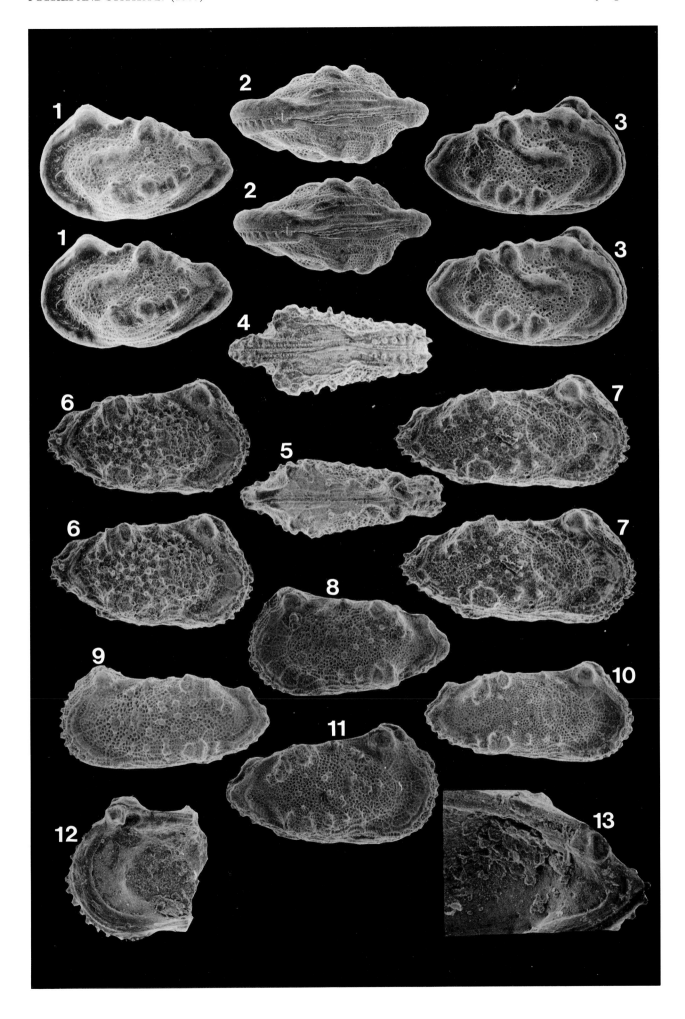

Plate 15

☐1. *Peloriops ziregensis?* (Bassoullet & Damotte). Left side of a female carapace, small specimen. Bordj Ghdir section, level A16. ×136. PMAL189. [Page 24.]

☐2. Same species and provenance. Right side of a male carapace, small specimen. ×136. PMAL190.

☐3. Same species?, somewhat uncertain provenance. Right side of a carapace, small specimen. ×136 PMAL191.

☐4. *Peloriops* sp. nov? Left side of a carapace. Bordj Ghdir section, level A10. ×77. PMAL192. [Page 25.]

☐5. Same species and provenance. Right side of a carapace. ×77. PMAL193.

☐6. *Peloriops* sp. Carapace in dorsal view. Qastel (Israel). ×102. PMIsr14. [Page 25.]

☐7. Same species and provenance. Left side of a carapace. ×98. PMIsr15.

☐8. Same species and provenance. Right side of a carapace. ×98. PMIsr16.

☐9. Same species and provenance. Carapace in ventral view. ×98. PMIsr17.

☐10. *Matronella* sp. Left side of a carapace. Bordj Ghdir section, level A2. ×64. PMAL194. [Page 25.]

☐11. Same individual. ×64.

☐12. *Curfsina?* sp. Right side of a carapace. Bordj Ghdir section, level A2. ×64. PMAL195. [Page 25.]

☐13. Same species and provenance. Right side of a carapace. ×64. PMAL196.

☐14. *Trachyleberidea?* sp. Right side of a carapace. Bordj Ghdir section, level A9. ×77. PMAL197. [Page 26.]

☐15. Same species; Bordj Ghdir section, level A8. Left side of a carapace. ×91. PMAL198.

☐16. Same species; Bordj Ghdir section, level A9. Right side of a male carapace. ×75. PMAL199.

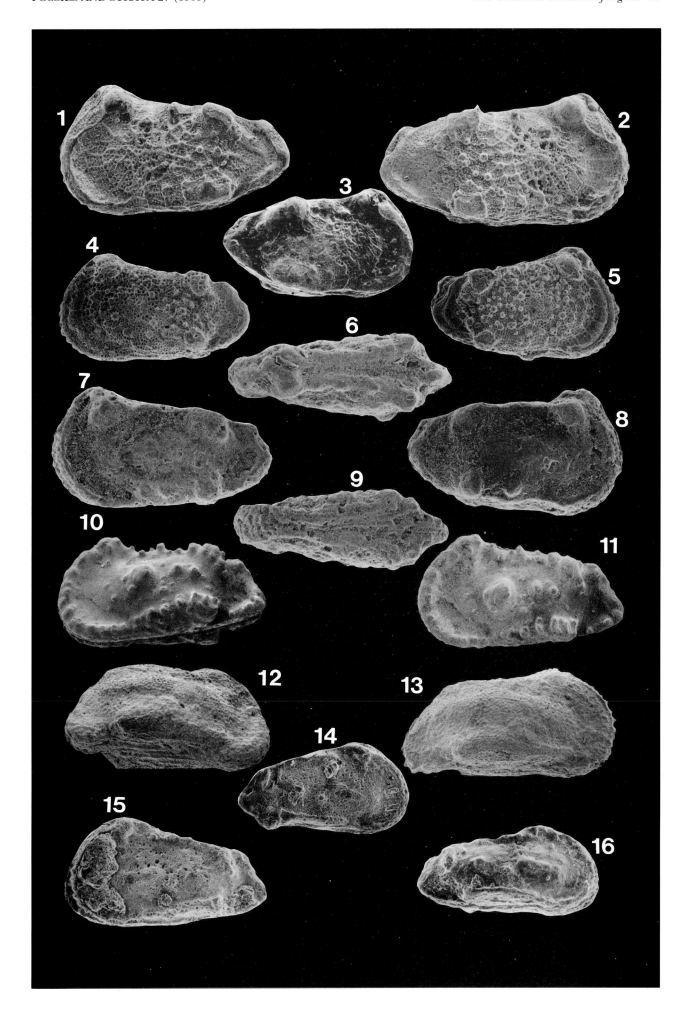

Plate 16

☐1. *Platycythereis?* sp. (juvenile). Left valve. Bordj Ghdir section, level A11. ×136. PMAL200. [Page 26.]

☐2. *Spinoleberis* sp. Left valve. Bordj Ghdir section, level A10. ×51. PMAL201. [Page 28.]

☐3. Same species and provenance. Carapace in ventral view. ×51. PMAL202.

☐4. *Spinoleberis kasserinensis* Bismuth & Saint-Marc. Left side of a carapace. Bordj Ghdir section, level A17. ×102. PMAL203. [Page 26.]

☐5. Same species and provenance. Right side of a carapace. ×102. PMAL204.

☐6. Same species from Djebel Semmama. Carapace in ventral view. ×102. PMTN42.

☐7. Same species and provenance. Left side of a carapace. ×102. PMTN43.

☐8. Same species and provenance. Right side of a carapace. ×102. PMTN44.

☐9. Same species and provenance. Carapace in dorsal view. ×136. PMTN45.

☐10. *Spinoleberis? yotvataensis?* Rosenfeld. Right side of a carapace. Somewhat uncertain provenance. ×68. PMAL205. [Page 27.]

☐11. Same species and provenance. Ventral view of a damaged carapace. ×64. PMAL206.

☐12. Same species and provenance. Right side of a carapace. ×75. PMAL207.

☐13. *Spinoleberis?* sp. Left side of a carapace. Somewhat uncertain provenance. ×77. PMAL208. [Page 27.]

☐14. Same species and provenance. Right side of a carapace. ×77. PMAL209.

☐15. *Brachycythere?* sp. Left side of a carapace. Somewhat uncertain provenance. ×55. PMAL210. [Page 28.]

☐16. *Protobuntonia?* sp. Left side of a carapace. Bordj Ghdir section, level A1. ×136. PMAL211. [Page 28.]

☐17. Same species and provenance. Right side of a carapace. ×136. PMAL212.

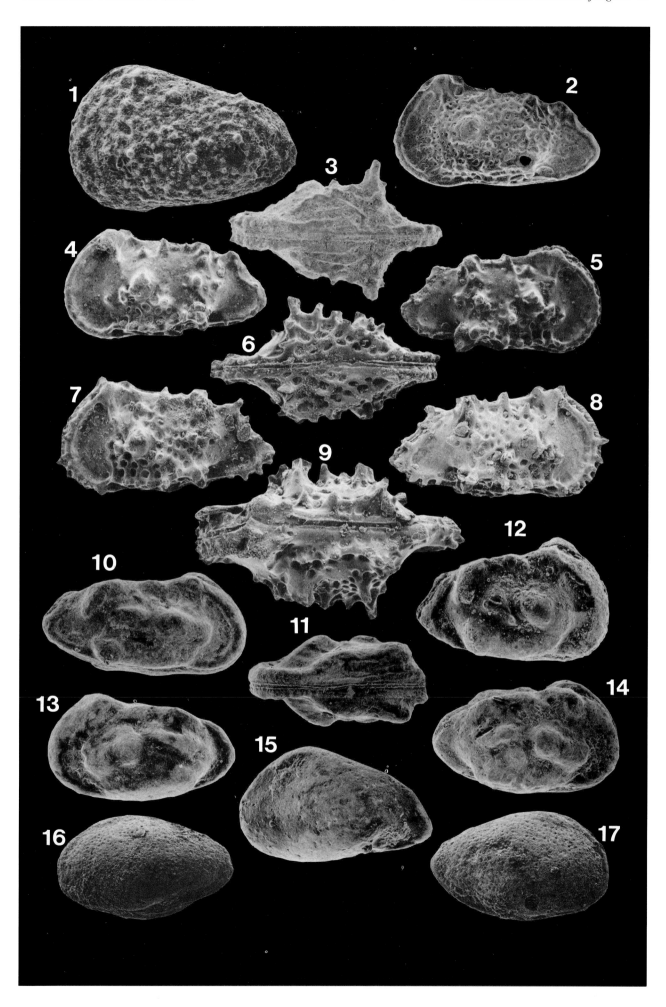

Plate 17

☐1. *Protobuntonia* sp. Left side of a female carapace. Somewhat uncertain provenance. ×66. PMAL213. [Page 28.]

☐2. Same species and provenance. Right side of a female carapace. ×66. PMAL214.

☐3. Same species and provenance. Right side of a male carapace. ×63. PMAL215.

☐4. *Limburgina* sp. Right side of a carapace (A-1?). Somewhat uncertain provenance. ×74. PMAL216. [Page 28.]

☐5. Same species and provenance. Right side of a carapace. ×73. PMAL217.

☐6. *'Dolocytheridea' polymorphica* sp.nov. Left female valve. Bordj Ghdir section, level A16. ×102. PMAL218. [Page 11.]

☐7. Same species and provenance. Left female valve. ×102. PMAL219.

☐8. Same species and provenance. Left female valve. ×102. PMAL220.

☐9. Same species and provenance. Left female valve. ×102. PMAL221.

☐10. Same species and provenance. Right male valve. ×102. PMAL222.

☐11. Same species and provenance. Left male valve. ×102. PMAL223.

☐12. Same species and provenance. Right male valve. ×102. PMAL224.

☐13. Same species and provenance. Right male valve. ×102. PMAL225.

☐14. Same species and provenance. Left male valve. ×102. PMAL226.

☐15. *Cytherella sulcata*? Van Veen. Right side of a male carapace. Bordj Ghdir section, level A13. ×68. PMAL227. [Page 8.]

☐16. Same species and provenance. Carapace in dorsal view. ×68. PMAL228.